HIT

国外优秀物理著作
原 版 系 列

范德瓦尔斯流体热力学的进展

Advances in Thermodynamics of the van der Waals Fluid

（英文）

[美] 大卫·C. 约翰斯顿（David C. Johnston） 著

HITP
哈爾濱工業大學出版社
HARBIN INSTITUTE OF TECHNOLOGY PRESS

黑版贸审字 08－2020－139 号

图书在版编目(CIP)数据

范德瓦尔斯流体热力学的进展＝Advances in Thermodynamics of the van der Waals Fluid：英文/(美)大卫・C. 约翰斯顿(David C. Johnston)著. —哈尔滨：哈尔滨工业大学出版社，2021. 7

ISBN 978-7-5603-9435-0

Ⅰ. ①范…　Ⅱ. ①大…　Ⅲ. ①流体力学－热力学－英文　Ⅳ. ①O351

中国版本图书馆 CIP 数据核字(2021)第 089119 号

策划编辑　刘培杰

责任编辑　聂兆慈　李兰静

封面设计　孙茵艾

出版发行　哈尔滨工业大学出版社

社　　址　哈尔滨市南岗区复华四道街 10 号　邮编 150006

传　　真　0451－86414749

网　　址　http://hitpress. hit. edu. cn

印　　刷　哈尔滨博奇印刷有限公司

开　　本　720 mm×1 000 mm　1/16　印张 7.75　字数 153 千字

版　　次　2021 年 7 月第 1 版　2021 年 7 月第 1 次印刷

书　　号　ISBN 978-7-5603-9435-0

定　　价　68.00 元

(如因印装质量问题影响阅读，我社负责调换)

Contents

Preface

This book evolved out of a course on thermodynamics and statistical mechanics that I taught at Iowa State University. One of the topics covered was phase transitions, and one of the types of phase transitions covered was the first-order liquid to gas transition described by the van der Waals mean-field theory of fluids. This theory was formulated by van der Waals in 1873 and, augmented by Maxwell in 1875, is the first theory that predicts a phase transition from interactions between particles. However, the thermodynamic properties of fluids (gases or liquids) derived from the van der Waals equation of state and free energy have not been thoroughly studied previously.

This book is a comprehensive exposition of the thermodynamic properties of the van der Waals fluid, which includes a review of past work together with presentation of my own recent extensive studies. The main goal of the book is to provide a graphical overview of the many interesting and diverse thermodynamic properties of the van der Waals fluid through plots of these properties versus various independent parameters. The data for these plots are obtained from formulas derived herein, some of which have previously appeared in the literature. Many results not amenable to graphical illustration are also included.

I hope that this book will be useful to instructors as a teaching resource and to students as a text supplement for thermodynamics and statistical mechanics courses as well as to others who are interested in the thermodynamics of the seminal van der Waals fluid.

David C Johnston
Ames, IA
May 27, 2014

Author biography

Dr David C Johnston

Dr David C Johnston is a Distinguished Professor in the Department of Physics and Astronomy of Iowa State University in Ames, Iowa. He received his BA and PhD degrees in Physics from the University of California at Santa Barbara and the University of California at San Diego, respectively. Prior to joining Iowa State University, he carried out research at the Corporate Research Laboratories of Exxon Research and Engineering Company in Annandale, NJ. His research area is experimental solid state physics, with an emphasis on the measurement and theoretical modeling of the electronic, magnetic, thermal and superconducting properties of solids. He is a Fellow of the American Physical Society and a former Divisional Associate Editor of the journal *Physical Review Letters*.

Advances in Thermodynamics of the van der Waals Fluid

David C Johnston

Chapter 1

Introduction

The van der Waals (vdW) mean-field theory for fluids containing interacting molecules was formulated by Johannes Diderik van der Waals in his PhD Thesis (Leiden) in 1873, for which he eventually won the Nobel Prize in Physics in 1910 'for his work on the equation of state for gases and liquids' [1–3][1]. An insightful discussion of the definition, usefulness as well as limitations of mean-field theories including that of van der Waals is given by Kadanoff [4].

The vdW fluid is the first, simplest and most widely known example of an interacting system of particles that exhibits a phase transition, in this case a first-order transition between liquid and gas (vapor) phases when the vdW theory of the pressure (p) versus volume (V) isotherms is complemented by the Maxwell construction [5] to define the regions of coexistence of gas and liquid. Despite its simplicity, the vdW fluid exhibits thermodynamic properties that are in semiquantitative agreement with the measured properties of real fluids. For these reasons the vdW equation of state and associated phase transition are presented in most thermodynamics and statistical mechanics courses and textbooks (see, e.g. [6–14]). Because of its richness and semiquantitive applicability to real fluids, the thermodynamics of the vdW fluid has enjoyed continued interest and development in modern times up to the present [15–22]. The critical exponents of several thermodynamic properties of the classical vdW fluid as the critical point is approached on specified thermodynamic paths are well known in the context of critical phenomena [23, 24].

Here, a critical and comprehensive study of the thermodynamic properties of the vdW fluid is presented for the first time. We show that despite its simplicity, the vdW equation of state leads to very interesting and remarkably diverse thermodynamic properties that are amazingly similar to those of real fluids. All thermodynamic

[1] [3] includes an introductory essay by the editor J S Rowlinson. The essay is an insightful retrospective on the van der Waals equation of state as presented in van der Waals' PhD thesis and on its antecedents, and on subsequent work by van der Waals and others on similar topics. The book also includes an edited English translation of van der Waals' PhD thesis.

doi:10.1088/978-1-627-05532-1ch1

properties are formulated in terms of reduced parameters that result in many laws of corresponding states, which by definition are the same for any fluid satisfying the assumptions of the vdW theory. These formulations allow the calculated thermodynamic properties to describe all vdW fluids. Since the purpose of this book is to review, clarify and elaborate on the thermodynamics of the vdW fluid specifically, we do not discuss the vast literature on equations of state for fluids that have higher accuracy than the vdW equation of state [25, 26]. Nor do we discuss in detail the ways in which the quantitative thermodynamics of real fluids deviate from the semiquantitative predictions of the vdW fluid, although several comparisons are made. Tests for the applicability of a particular equation of state to a particular temperature range of a fluid are discussed in [26] and [27].

In chapter 2, the nomenclature and definitions of the thermodynamics functions and properties used here are briefly discussed along with the well-known properties of the ideal gas for reference. The vdW molecular interaction parameters a and b are discussed in chapter 3 in terms of the Lennard-Jones potential where the ratio a/b is shown to be a fixed value for a particular vdW fluid which is determined by the depth of the Lennard-Jones potential well for that fluid. We only consider here molecules without internal degrees of freedom. The Helmholtz free energy F, the critical pressure p_c, temperature T_c and volume V_c and critical compressibility factor Z_c are defined in terms of a and b in chapter 4, which then allows the values of a and b for a particular fluid to be determined from the measured values of p_c and T_c for the fluid. The entropy S, internal energy U and heat capacity at constant volume C_V for the single-phase vdW fluid are written in terms of a, b, the volume V occupied by the fluid and the number N of molecules in section 4.3, the pressure and enthalpy H in section 4.4 and the chemical potential μ in section 4.5. The vdW equation of state is written in terms of dimensionless reduced variables in chapter 5 and the definition of the laws of corresponding states is discussed, together with plots of p versus V and p versus number density n isotherms, V versus T isobars and μ versus V isotherms, where the reduced variables are used; these plots contain unstable and metastable regions that are removed later when the equilibrium liquid–gas coexistence regions are determined.

There has been much discussion and disagreement about the influence of the vdW molecular interaction parameters a and/or b on the pressure of a vdW gas compared to that of an ideal gas at the same volume and temperature. This topic is quantitatively discussed in section 5.4 where it is shown that the pressure of a vdW gas can either increase or decrease compared to that of an ideal gas depending on the volume and temperature of the gas. A related topic is the Boyle temperature T_B at which the 'compression factor' $Z \equiv pV/(Nk_BT_B)$ is the same as for the ideal gas as discussed in section 5.5, where k_B is Boltzmann's constant. It is sometimes stated that at the Boyle temperature, the properties of a gas are the same as for the ideal gas; we show that this inference is incorrect for the vdW gas because even at this temperature other thermodynamic properties are not the same as those of an ideal gas. In section 5.6, the thermodynamic variables and functions are written in terms of dimensionless reduced variables, which are used in the remainder of the book.

The equilibrium values of the pressure and coexisting liquid and gas volumes are calculated in chapter 6 using traditional methods such as the Maxwell construction and the equilibrium p–V, T–V and p–T isotherms and phase diagrams including equilibrium gas–liquid coexistence regions are presented. The lever rule is derived in our notation which is needed later to calculate $C_V(T)$ along isochores in the liquid–gas coexistence region.

An important and very useful advance in calculating the gas–liquid coexistence curve in the p–T plane and associated properties of the vdW fluid was presented by Lekner in 1982 [15], who formulated a parametric solution in terms of the entropy difference per molecule between the gas and liquid phases. Lekner's results were extended by Berberan-Santos *et al* in 2008 [16]. In chapter 7, Lekner's results are explained and further extended for the full temperature region up to the critical temperature and the limiting behaviors of properties associated with the coexistence curve for $T \to 0$ and $T \to T_c$ are also calculated. In the topics and regions of overlap these results agree with the previous ones [15, 16]. In sections 7.3 and 7.4, the coexisting liquid and gas densities, the difference between them which is the order parameter for the gas–liquid phase transition, the T–n phase diagram, and the latent heat and entropy of vaporization utilizing Lekner's parametrization are calculated and plotted.

In section 7.5, the C_V is calculated versus temperature along isochoric paths within the liquid–gas coexistence region, apparently for the first time. We use Lekner's parametric solution to obtain numerically exact results. The calculation of C_V along the path of the critical isochore allows the critical behavior for $T \to T_c^-$ to be determined and it is found to be in agreement with previous results. At first sight, since the C_V of pure gas and liquid vdW phases are independent of T, one might expect C_V in the coexistence region to also be independent of T. We show that this is not the case. The T dependence arises from the T dependences of the entropy difference per molecule between the liquid and gas phases and of the mole fractions of coexisting liquid and gas phases.

Tables of calculated values of parameters and properties obtained using both the conventional and Lekner parameterizations of the coexisting liquid and gas phases are given in appendix A. Analytic expressions for the discontinuities in the isothermal compressibility κ_T, thermal volume expansion coefficient α and heat capacity at constant pressure C_p on crossing the coexistence curve in the p–T plane derived using Lekner's parameterization are given in appendix B. An analytic expression for C_V along the critical isochore within the coexistence region derived using Lekner's parameterization is given in appendix C.

The order parameter for the liquid–gas transition is the difference in density between the liquid and gas phases, which goes to zero on approaching the critical point from below. Critical exponents and amplitudes for the vdW fluid are calculated in chapter 8 for the order parameter, C_V, κ_T, α and C_p, where their values can depend on the path of approach to the critical point. The critical amplitudes are expressed in terms of the universal reduced parameters used throughout this book.

In chapter 9, the maximum hysteresis in the transition temperature on heating (superheating) and cooling (supercooling) through the equilibrium first-order liquid–gas phase transition temperature at constant pressure is evaluated and plotted versus temperature. Such hysteresis is generic to first-order phase transitions, but to our knowledge has not been quantitatively calculated previously for the vdW fluid. A first-order phase transition is one where the order parameter for the transition (the difference between the liquid and gas densities in the case of fluids) changes discontinuously upon heating or cooling through the equilibrium transition temperature, such as the boiling of liquid water into steam at the boiling temperature of 100 °C at one atmosphere pressure. A latent heat of vaporization is associated with such fluid transitions.

Numerical calculations of the isothermal number density fluctuation $N_{\rm fluct}$, $\kappa_{\rm T}$, α and $C_{\rm p}$ versus n are presented for $T > T_{\rm c}$ in chapter 10. In this supercritical region, these thermodynamic functions continuously change their temperature dependences toward their divergent critical behaviors for $T \to T_{\rm c}^{+}$. 'Quasicritical' curves versus T and n that are the loci (ridges) of the temperature of zero curvature for $N_{\rm fluct}$ and of the peak positions of $\kappa_{\rm T}$, α and $C_{\rm p}$ versus T and n are studied in this chapter.

Additional numerical calculations of n, $\kappa_{\rm T}$, α and $C_{\rm p}$ versus temperature at constant pressure for $p > p_{\rm c}$, $p = p_{\rm c}$ and $p < p_{\rm c}$ are presented in section 10.2. The fitted critical exponents and amplitudes for $\kappa_{\rm T}$, α and $C_{\rm p}$ for the critical isobar $p = p_{\rm c}$ are found to agree with the corresponding behaviors predicted analytically in chapter 8. The discontinuities in the calculated $\kappa_{\rm T}$, α and $C_{\rm p}$ on crossing the coexistence curve in the p–T plane at constant pressure with $p < p_{\rm c}$ are also shown to be in agreement with the analytic predictions in appendix B noted above. For temperatures $T > T_{\rm c}$, the data show quasicritical ridges as defined above, and these ridges for n, $\kappa_{\rm T}$, α and $C_{\rm p}$ asymptotically approach each other for $T \to T_{\rm c}^{+}$. The asymptotic behavior is called the 'Widom line', [28] and we give the equation for it [22].

Cooling the vdW gas by adiabatic free expansion and cooling and/or liquifying the vdW gas by Joule–Thomson expansion are discussed in chapter 11, where the conditions for liquification of a vdW gas on passing through a throttle are presented. The process of Joule–Thomson expansion resulting in liquifaction of the vdW gas has been little discussed previously. An analytic equation for the inversion curve associated with the Joule–Thomson expansion of the vdW fluid is derived and found to be consistent with that previously reported by Le Vent in 2001 [18].

References

[1] van der Waals J D 1873 *PhD thesis* Leiden University
van der Waals J D 1988 *On the Continuity of the Gaseous and Liquid States* (Mineola, NY: Dover) (Engl. transl.)

[2] van der Waals J D 1910 *Nobel Lecture* December 12

[3] Rowlinson J S (ed) 1988 *J D van der Waals: On the Continuity of the Gaseous and Liquid States* (Amsterdam: North-Holland)

[4] Kadanoff L P 2009 More is the same; phase transitions and mean field theories *J. Stat. Phys.* **137** 777

[5] Clerk-Maxwell J 1875 On the dynamical evidence of the molecular constitution of bodies *Nature* **11** 357
[6] Reif F 1965 *Fundamentals of Statistical and Thermal Physics* (New York: McGraw-Hill)
[7] Kittel C and Kroemer H 1980 *Thermal Physics* 2nd edn (New York: Freeman)
[8] Schroeder D V 2000 *An Introduction to Thermal Physics* (San Francisco, CA: Addison Wesley Longman)
[9] Reichl L E 2009 *A Modern Course in Statistical Physics* 3rd edn (Weinheim: Wiley-VCH)
[10] Callen H B 1960 *Thermodynamics* (New York: Wiley)
[11] Landau L D and Lifshitz E M 1969 *Statistical Physics* 2nd edn (Oxford: Pergamon)
[12] Huang K 1963 *Statistical Mechanics* (New York: Wiley)
[13] Zemansky M W and Dittman R H 1997 *Heat and Thermodynamics* 7th edn (New York: McGraw-Hill)
[14] Rowlinson J S and Swinton F L 1982 *Liquids and Liquid Mixtures* (London: Butterworth)
[15] Lekner J 1982 Parametric solution of the van der Waals liquid–vapor coexistence curve *Am. J. Phys.* **50** 161
[16] Berberan-Santos M N, Bodunov E N and Pogliani L 2008 The van der Waals equation: analytical and approximate solutions *J. Math. Chem.* **43** 1437
[17] Swendsen R H 2013 Using computation to teach the properties of the van der Waals fluid *Am. J. Phys.* **81** 776
[18] Le Vent S 2001 A summary of the properties of van der Waals fluids *Int. J. Mech. Eng. Educ.* **29** 257
[19] Tuttle E R 1975 The cohesion term in van der Waals's equation of state *Am. J. Phys.* **43** 644
[20] Kwok C K and Tilley D R 1979 A review of some thermodynamic properties of the van der Waals gas *Phys. Educ.* **14** 422
[21] Nishikawa K, Kusano K, Arai A A and Morita T 2003 Density fluctuation of a van der Waals fluid in supercritical state *J. Chem. Phys.* **118** 1341
[22] Brazhkin V V and Ryzhov V N 2011 van der Waals supercritical fluid: exact formulas for special lines *J. Chem. Phys.* **135** 084503
[23] Stanley H E 1971 *Introduction to Phase Transitions and Critical Phenomena* (New York: Oxford Science)
[24] Heller P 1967 Experimental investigations of critical phenomena *Rep. Prog. Phys.* **30** 731
[25] Sengers J V, Kayser R F, Peters C J and White B H (ed) 2000 *Equations of State for Fluids and Fluid Mixtures, IUPAC Monograph Series in Experimental Thermodynamics* vol 5 (Amsterdam: Elsevier)
[26] Deiters U K and Kraska T 2012 *High-Pressure Fluid Phase Equilibria; Supercritical Fluid Science and Technology* vol 2 (Amsterdam: Elsevier) chapter 7
[27] Deiters U K and de Reuck K M 1997 Guidelines for publication of equations of state–I. Pure fluids *Pure Appl. Chem.* **69** 1237
[28] Xu L, Kumar P, Buldyrev S V, Chen S-H, Poole P H, Sciortino F and Stanley H E 2005 Relation between the Widom line and the dynamic crossover in systems with a liquid–liquid phase transition *Proc. Natl Acad. Sci. USA* **102** 16558

David C Johnston

Chapter 2

Background and nomenclature: the ideal gas

The following notations, definitions and relations mostly follow [1]. An ideal gas is defined as a gas of noninteracting particles in the classical regime where the number density of the gas is small. In this case one obtains an equation of state called the ideal gas law

$$pV = Nk_{\mathrm{B}}T = N\tau, \tag{2.1a}$$

where throughout this book we use the notation

$$\tau \equiv k_{\mathrm{B}}T. \tag{2.1b}$$

For an ideal gas containing molecules with no internal degrees of freedom, the Helmholtz free energy F is

$$F(\tau,\ V,\ N) = -N\tau\left\{\ln\left[\frac{n_Q V}{N}\right] + 1\right\}, \tag{2.2a}$$

where the 'quantum concentration' n_Q is given by

$$n_Q = \left(\frac{m\tau}{2\pi\hbar^2}\right)^{3/2}, \tag{2.2b}$$

m is the mass of a molecule and $\hbar$ is Planck's constant divided by 2π. Other authors instead use an expression containing the 'thermal wavelength' λ_{T} defined by $\lambda_{\mathrm{T}}^3 = 1/n_Q$.

The entropy S is

$$\frac{S}{k_{\mathrm{B}}} = -\left(\frac{\partial F}{\partial \tau}\right)_{V,N} = N\left[\ln\left(\frac{n_Q V}{N}\right) + \frac{5}{2}\right]. \tag{2.3}$$

doi:10.1088/978-1-627-05532-1ch2

This equation is known as the Sackur–Tetrode equation. The internal energy U is

$$U = F + TS = \frac{3}{2}N\tau. \tag{2.4}$$

The internal energy of an ideal gas only depends on the temperature and not on the volume because, by definition, the molecules in an ideal gas do not see each other and therefore the distance between the molecules and hence the volume are not relevant. The heat capacity at constant volume C_V is given by

$$C_\mathrm{V} = k_\mathrm{B}\left(\frac{\partial U}{\partial \tau}\right)_{V,N} = \frac{3}{2}Nk_\mathrm{B} \tag{2.5}$$

and the enthalpy H by

$$H = U + pV = \frac{5}{2}N\tau = \frac{5}{2}Nk_\mathrm{B}T. \tag{2.6}$$

The isothermal compressibility κ_T is

$$\kappa_\mathrm{T} = -\frac{1}{V}\left(\frac{\partial V}{\partial p}\right)_T = \frac{1}{p}, \tag{2.7}$$

and the volume thermal expansion coefficient α is

$$\alpha = \frac{1}{V}\left(\frac{\partial V}{\partial T}\right)_p = \frac{Nk_\mathrm{B}}{pV} = \frac{1}{T}. \tag{2.8}$$

The heat capacity at constant pressure C_p is given by [2]

$$C_\mathrm{p} = C_\mathrm{V} + \frac{TV\alpha^2}{\kappa_\mathrm{T}} = \frac{3}{2}Nk_\mathrm{B} + \frac{pV}{T} = \frac{5}{2}Nk_\mathrm{B}. \tag{2.9}$$

Alternatively,

$$C_\mathrm{p} = \left(\frac{\partial H}{\partial T}\right)_p \tag{2.10}$$

gives the same result.

The chemical potential μ is

$$\mu = \left(\frac{\partial F}{\partial N}\right)_{\tau,V} = \tau\ln\left(\frac{N}{n_Q V}\right) = \tau\ln\left(\frac{p}{n_Q\tau}\right), \tag{2.11}$$

where to obtain the last equality we used the ideal gas law equation (2.1a). The Gibbs free energy G written in terms of its natural variables N, p and T or τ is

$$G(N, p, \tau) = N\mu(p, \tau), \tag{2.12a}$$

where the differential of G is

$$dG = -S\,dT + V\,dp + \mu\,dN. \tag{2.12b}$$

References

[1] Kittel C and Kroemer H 1980 *Thermal Physics* 2nd edn (New York: Freeman)
[2] Reif F 1965 *Fundamentals of Statistical and Thermal Physics* (New York: McGraw-Hill)

Chapter 3

van der Waals intermolecular interaction parameters

Interactions between neutral molecules or atoms with a center of mass separation r are often approximated by the so-called Lennard-Jones potential energy $\phi_{\rm LJ}$, given by

$$\phi_{\rm LJ}(r) = 4\phi_{\rm min}\left[\left(\frac{r_0}{r}\right)^{12} - \left(\frac{r_0}{r}\right)^{6}\right], \tag{3.1}$$

where the first term is a short-range repulsive interaction and the second term is a longer-range attractive interaction. A plot of $\phi_{\rm LJ}/\phi_{\rm min}$ versus r/r_0 is shown in figure 3.1. The value $r = r_0$ corresponds to $\phi_{\rm LJ} = 0$, and the minimum value of $\phi_{\rm LJ}$ is $\phi_{\rm LJ}/\phi_{\rm min} = -1$ at

$$r_{\rm min}/r_0 = 2^{1/6} \approx 1.122. \tag{3.2}$$

By the definition of potential energy, the force of a molecule on a neighboring molecule in the radial direction from the first molecule is $F_r = -{\rm d}\phi_{\rm LJ}/{\rm d}r$, which is positive (repulsive) for $r < r_{\rm min}$ and negative (attractive) for $r > r_{\rm min}$.

In the vdW theory of a fluid (gas and/or liquid) discussed here, one ignores possible internal degrees of freedom of the molecules and assumes that the interatomic distance between molecules cannot be smaller than a molecular diameter, a situation called a 'hard-core repulsion' where two molecules cannot overlap. Therefore the minimum intermolecular distance from center to center is equal to the diameter d of one molecule. In terms of the Lennard-Jones potential, we set

$$d = r_{\rm min}, \tag{3.3}$$

rather than $d = r_0$, because the Lennard-Jones interaction between two molecules is repulsive out to a separation of $r_{\rm min}$ as shown in figure 3.1. In the vdW theory, the effective volume of a molecule ('excluded volume' or 'co-volume') is denoted by

doi:10.1088/978-1-627-05532-1ch3

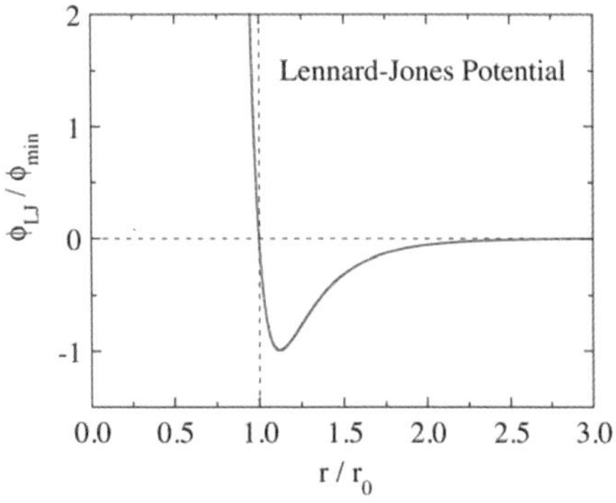

Figure 3.1. Lennard-Jones potential energy ϕ_{LJ} between two molecules versus the distance r between the centers of the two molecules as predicted by equation (3.1). The depth of the potential well is ϕ_{min} and the zero of the potential energy is at $r = r_0$.

the variable b, so the free volume available for the molecules to move in is $V - Nb$. Thus in the free energy of the ideal gas in equation (2.2a) one makes the substitution

$$V \rightarrow V - Nb. \tag{3.4}$$

In terms of the Lennard-Jones potential, we set

$$b \equiv d^3 = r_{min}^3 = \sqrt{2}\, r_0^3, \tag{3.5}$$

where $d = r_{min}$ is a measure of the hard-core diameter of a molecule and we have used equation (3.2). For $r > d$, the force between the molecules in a vdW fluid is assumed to be attractive and the strength of the attraction depends on the distance between the molecules. In terms of the Lennard-Jones potential this occurs for $r > r_{min}$ according to equation (3.1) and figure 3.1.

One takes into account the attractive part of the interaction for a general potential $\phi(r)$ in an average way as follows, which is a 'mean-field' approximation where one ignores local fluctuations in the number density of molecules and short-range correlations between their positions. The number density of molecules is N/V. The number $\mathrm{d}N$ of molecules that are at a distance between r and $r + \mathrm{d}r$ from the central molecule is $\mathrm{d}N = (N/V)\mathrm{d}V$, where an increment of volume a distance r from the center of the central molecule is $\mathrm{d}V = 4\pi r^2 \mathrm{d}r$. Thus the total average attractive potential energy summed over these molecules, ϕ_{ave}, is

$$\phi_{ave} = \left(\frac{N}{V}\right)\frac{1}{2}\int_{r_{min}}^{\infty} \phi(r)\mathrm{d}V = \left(\frac{N}{V}\right)\frac{4\pi}{2}\int_{r_{min}}^{\infty} \phi(r) r^2 \mathrm{d}r, \tag{3.6}$$

where the prefactor of 1/2 arises because the potential energy of interaction between a molecule and a neighboring molecule is shared equally between them. In the vdW theory, one writes the average potential energy per molecule as

$$\phi_{ave} = -\left(\frac{N}{V}\right)a \tag{3.7}$$

where the parameter $a \geqslant 0$ is an average value of the potential energy per unit concentration, given here using equation (3.6) by

$$a = -2\pi \int_{r_{\min}}^{\infty} \phi(r) r^2 \mathrm{d}r. \tag{3.8}$$

One can obtain an expression for a in terms of the Lennard-Jones potential $\phi_{\mathrm{LJ}}(r)$. Substituting the Lennard-Jones potential in equation (3.1) into (3.8), one has

$$a = -8\pi\phi_{\min} \int_{r_{\min}}^{\infty} \left[\left(\frac{r_0}{r}\right)^{12} - \left(\frac{r_0}{r}\right)^{6} \right] r^2 \mathrm{d}r. \tag{3.9}$$

Changing variables to $x = r/r_0$ and using equation (3.2) gives

$$a = -8\pi\phi_{\min} r_0^3 \int_{2^{1/6}}^{\infty} \left(\frac{1}{x^{10}} - \frac{1}{x^4} \right) \mathrm{d}x. \tag{3.10}$$

The integral is $-5/(18\sqrt{2})$, yielding

$$a = \frac{20\pi r_0^3 \phi_{\min}}{9\sqrt{2}}. \tag{3.11}$$

From equations (3.5) and (3.11), a and b per molecule are related to each other according to

$$\frac{a}{b} = \frac{10\pi\phi_{\min}}{9}. \tag{3.12}$$

This illustrates the important feature that the ratio a/b for a given vdW fluid is a fixed value that depends on the intermolecular potential function.

Chapter 4

Thermodynamic variables and properties in terms of the van der Waals interaction parameters

4.1 Helmholtz free energy

The change in the internal energy due to the attractive part of the intermolecular interaction is the potential energy $N\phi_{\text{ave}}$ and from equation (3.7) one obtains

$$\Delta U = N\phi_{\text{ave}} = -\frac{N^2 a}{V}. \tag{4.1}$$

When one smoothly turns on interactions in a thought experiment, effectively one is doing work on the system and this does not transfer thermal energy. Therefore the potential energy represented by the parameter a introduces no entropy change and hence the change in the free energy is $\Delta F = \Delta U - \Delta(TS) = \Delta U$. The attractive part of the intermolecular potential energy that results in a change in F compared to the free energy of the ideal gas is then given by

$$F \to F + \Delta U = F - \frac{N^2 a}{V}. \tag{4.2}$$

Making the changes in equations (3.4) and (4.2) to the free energy of the ideal gas in equation (2.2a) gives the free energy of the vdW fluid as

$$F(\tau, V, N) = -N\tau\left\{\ln\left[\frac{n_Q(V - Nb)}{N}\right] + 1\right\} - \frac{N^2 a}{V}. \tag{4.3}$$

This is a quantum mechanical expression because $\hbar$ is present in n_Q. However, we will see that the thermodynamic properties of the vdW fluid are classical, where

doi:10.1088/978-1-627-05532-1ch4

$\hbar$ does not appear in the final calculations. In the limit $N/V \to 0$ or equivalently a, $b \to 0$, the Helmholtz free energy becomes that of the ideal gas in equation (2.2*a*).

4.2 Critical pressure, temperature and volume

The critical pressure p_c, critical volume V_c and critical temperature $\tau_c \equiv k_B T_c$ define the critical point of the vdW fluid as discussed later. These are given in terms of the parameters a, b and N as

$$p_c = \frac{a}{27b^2} \qquad V_c = 3Nb \qquad \tau_c = \frac{8a}{27b}. \tag{4.4a}$$

The product of the first two expressions gives an energy scale

$$p_c V_c = \frac{Na}{9b} = \frac{3N\tau_c}{8} = \frac{3Nk_B T_c}{8}, \tag{4.4b}$$

yielding the universal ratio called the critical 'compression factor' Z_c as

$$Z_c \equiv \frac{p_c V_c}{N\tau_c} = \frac{3}{8}. \tag{4.4c}$$

The critical temperatures, pressures and volumes of representative gases are shown in table 4.1. One sees from the table that the experimental values of Z_c are ~30% smaller that the value of 3/8 predicted by the vdW theory in equation (4.4*c*), indicating that the theory does not accurately describe real gases. One can solve equations (4.4*a*) for a, b and N in terms of the critical variables, yielding

$$a = \frac{27\tau_c^2}{64p_c} = 27b^2 p_c \qquad b = \frac{\tau_c}{8p_c} \qquad N = \frac{8p_c V_c}{3\tau_c}. \tag{4.4d}$$

Shown in table 4.1 are the vdW parameters a and b per molecule derived from the measured values of T_c and p_c [1] using the first two of equations (4.4*d*). The listed values of a and b are expressed in units associated with a molecule such as eV and Å, which are more physically relevant to the molecules composing the fluid than the common units of these quantities, which are, e.g., bar (L mol^{-1})^2 and L mol^{-1}, respectively. Thus the parameter b is the excluded volume per molecule expressed in units of Å^3, from which the effective diameter per molecule d in Å is obtained here as $d = b^{1/3}$ as shown in the table. From the second of equations (4.4*a*), the critical volume per molecule is $V_c/N = 3b$, which is only a factor of three larger than the excluded volume of a molecule itself. Shown in the last column of table 4.1 is the effective Lennard-Jones intermolecular potential well depth $\phi_{\min}$ in figure 3.1 calculated from a and b using equation (3.12). The values of $\phi_{\min}$ are seen to be smallest for He and H_2 and largest for H_2O and the heavy alkanes.

The dependences of a and b on the molecular weight (MW) of the various gases in table 4.1 are shown in figures 4.1(a) and 4.1(b), respectively. For fluids of one of the types shown, the two parameters a and b do not increase monotonically with molecular weight except for the alkanes.

Table 4.1. Experimental data for representative gases obtained from [1]. Shown are the molecular weight (MW), critical temperature T_c, critical pressure p_c, critical volume V_c and the dimensionless critical compression factor $Z_c \equiv p_c V_c/(RT_c)$ where R is the molar gas constant. The value of Z_c predicted by the vdW theory is $Z_c = 3/8 = 0.375$ according to equation (4.4*c*), which is ~30% larger than the observed factors listed in the table. Also shown are the vdW parameters a and b per molecule derived from T_c and p_c using the first two of equations (4.4*d*), where a is a mean-field measure of the attractive force between two molecules and b is the excluded volume per molecule due to the molecular hard cores. A measure of the vdW hard-core molecular diameter is defined here as $d \equiv b^{1/3}$. Assuming a Lennard-Jones potential between molecules, the depth ϕ_{min} of the potential well in figure 3.1 is calculated from a/b using equation (3.12).

Gas Name	Formula	MW (g mol^{-1})	T_c (K)	p_c (kPa)	V_c (cm^3 mol^{-1})	Z_c	a (eV Å^3)	b (Å^3)	d (Å)	ϕ_{min} (meV)
Noble gases										
Helium	He	4.0030	5.1953	227.46	57	0.300	0.05956	39.418	3.4033	0.433
Neon	Ne	20.183	44.490	2678.6	42	0.304	0.37090	28.665	3.0604	3.707
Argon	Ar	39.948	150.69	4863	75	0.291	2.344	53.48	3.768	12.56
Krypton	Kr	83.800	209.48	5525	91	0.289	3.987	65.43	4.030	17.45
Xenon	Xe	131.30	289.73	5842	118	0.286	7.212	85.59	4.407	24.14
Diatomic gases										
Hydrogen	H_2	2.0160	33.140	1296.4	65	0.306	0.42521	44.117	3.5335	2.761
Hydrogen fluoride	HF	20.006	461.00	6480	69	0.117	16.46	122.8	4.970	38.41
Nitrogen	N_2	28.014	126.19	3390	90	0.291	2.358	64.24	4.005	10.51
Carbon monoxide	CO	28.010	132.86	3494	93	0.294	2.536	65.62	4.034	11.07
Nitric oxide	NO	30.010	180.00	6480	58	0.251	2.510	47.94	3.633	15.00
Oxygen	O_2	32.000	154.58	5043	73	0.286	2.378	52.90	3.754	12.88
Hydrogen chloride	HCl	36.461	324.70	8310	81	0.249	6.368	67.43	4.070	27.05
Fluorine	F_2	37.997	144.41	5172.4	66	0.284	2.024	48.184	3.6389	12.03
Chlorine	Cl_2	70.910	417.00	7991	123	0.284	10.92	90.06	4.482	34.74
Polyatomic gases										
Ammonia	NH_3	17.031	405.56	11357	69.9	0.235	7.2692	61.629	3.9500	33.79
Water	H_2O	18.015	647.10	22060	56	0.230	9.5273	50.624	3.6993	53.92
Carbon dioxide	CO_2	44.010	304.13	7375	94	0.274	6.295	71.17	4.144	25.34

(Continued)

Table 4.1. (*Continued*)

Gas Name	Formula	MW (g mol^{-1})	T_c (K)	p_c (kPa)	V_c (cm^3 mol^{-1})	Z_c	a (eV Å^3)	b (Å^3)	d (Å)	ϕ_{min} (meV)
Nitrous oxide	N_2O	44.013	309.52	7245	97	0.273	6.637	73.73	4.193	25.79
Carbon oxysulfide	COS	60.074	375.00	5880	137	0.258	12.00	110.1	4.792	31.24
Alkanes										
Methane	CH_4	16.043	190.56	4600	99	0.287	3.962	71.49	4.150	15.88
Ethane	C_2H_6	30.070	305.36	4880	146	0.281	9.591	108.0	4.762	25.44
Propane	C_3H_8	44.097	369.9	4250	199	0.275	16.16	150.2	5.316	30.82
Butane	C_4H_{10}	55.124	425.2	3790	257	0.276	23.94	193.6	5.785	35.43
Pentane	C_5H_{12}	72.151	469.7	3370	310	0.268	32.86	240.5	6.219	39.13
Hexane	C_6H_{14}	86.178	507.5	3030	366	0.263	42.67	289.1	6.612	42.28
Heptane	C_7H_{16}	100.21	540.1	2740	428	0.261	53.44	340.2	6.981	45.00

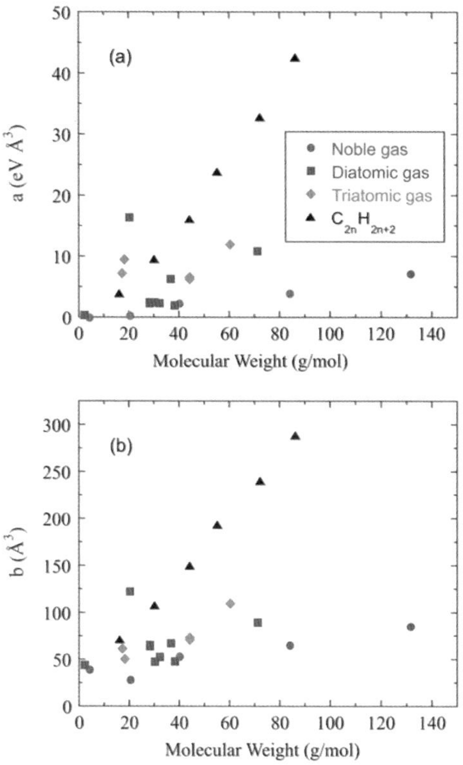

Figure 4.1. vdW parameters per molecule (*a*) a and (*b*) b versus the molecular weight of the noble (monatomic) gases, diatomic, triatomic/polyatomic and alkane (C_nH_{2n+2}) gases that are listed in table 4.1.

4.3 Entropy, internal energy and heat capacity at constant volume

The entropy of the vdW fluid is calculated using equation (4.3) to be

$$\frac{S}{k_{\mathrm{B}}} = -\left(\frac{\partial F}{\partial \tau}\right)_{V,N} = N\left\{\ln\left[\frac{n_Q(V - Nb)}{N}\right] + \frac{5}{2}\right\}, \tag{4.5}$$

which is smaller than that of the ideal gas in equation (2.3) because the entropy scales with the free volume, which is smaller in the vdW fluid. In the limits $V \to \infty$ or $b \to 0$, equation (4.5) becomes identical to (2.3).

The internal energy is obtained using equations (4.3) and (4.5) as

$$U = F + TS = \frac{3}{2}N\tau - \frac{N^2 a}{V}, \tag{4.6}$$

which is lower than that of the ideal gas in equation (2.4) by the attractive potential energy in the second term on the right. However, because the interaction parameter a is independent of temperature, it does not contribute to the temperature dependence of U given by the first term on the right-hand side of equation (4.6) which is the same as for the ideal gas in equation (2.4).

Since the temperature dependence of the internal energy of the vdW fluid is the same as for the ideal gas, the heat capacity at constant volume for a single-phase vdW fluid (liquid, gas or undifferentiated fluid, see below) is

$$C_{\mathrm{V}} = \left(\frac{\partial U}{\partial T}\right)_{V,N} = \frac{3}{2}Nk_{\mathrm{B}}, \tag{4.7}$$

which is the same as for the ideal gas in equation (2.5). This heat capacity is independent of T, so the vdW fluid is in the classical limit of a quantum Fermi or Bose gas. Furthermore, the forms of the thermodynamic functions are the same for the pure gas, pure liquid and undifferentiated fluid phases, which only differ in the temperature, pressure and volume regions in which they occur. Therefore, in particular, the pure gas, liquid and fluid phases discussed below have the same constant value of C_{V}. However, a remarkably different situation occurs in the coexistence region of gas and liquid, where we find in section 7.5 that C_{V} is temperature-dependent.

4.4 Pressure and enthalpy

The pressure p is obtained from the free energy in equation (4.3) as

$$p = -\left(\frac{\partial F}{\partial V}\right)_{\tau,N} = \frac{N\tau}{V - Nb} - \frac{N^2 a}{V^2}. \tag{4.8}$$

As discussed above, the volume Nb is the excluded volume of the incompressible molecules and $V - Nb$ is the free volume in which the molecules can move. With decreasing volume V, the pressure diverges when $Nb = V$ because then all of the volume is occupied by the total excluded volume of the molecules themselves and the incompressible hard cores of the molecules are touching. Therefore the minimum possible volume of the system is $V_{\min} = Nb$. Hence the first term on the right-hand side of equation (4.8) is always positive and the second term negative. The competition between these two terms in changing the pressure of the gas, compared to that of an ideal gas at the same temperature and volume, is quantitatively discussed in section 5.4 below.

Plots of $p(V)$ at constant temperature using equation (4.8) have the shapes shown in figure 5.1 below. At the critical point $\tau = \tau_{\mathrm{c}}$, $p = p_{\mathrm{c}}$ and $V = V_{\mathrm{c}}$, $p(V)$ shows an inflection point where the slope $(\partial p/\partial V)_\tau$ and the curvature $(\partial^2 p/\partial V^2)_\tau$ are both zero. From these two conditions one can solve for the the critical temperature τ_{c} and pressure p_{c} in terms of the vdW parameters a and b, and then from the equation of state one can solve for the critical volume V_{c} in terms of a, b and N as given above in equations (4.4a).

Using equations (4.6) and (4.8), the enthalpy is

$$H = U + pV = N\left(\frac{3\tau}{2} + \frac{\tau V}{V - Nb} - \frac{2Na}{V}\right). \tag{4.9}$$

In the limit of infinite volume V or vanishing interaction parameters a and b, one obtains equation (2.6) for the enthalpy of the ideal gas.

4.5 Chemical potential

The vdW fluid contains attractive interactions and one therefore expects that it may liquify at sufficiently low temperature and/or sufficiently high pressure. The liquid (l) phase is more stable than the gas (g) phase when the liquid and gas chemical potentials satisfy $\mu_l < \mu_g$, or equivalently, when the Gibbs free energy satisfies $G_l < G_g$ and the gas phase is more stable when $\mu_l > \mu_g$ or $G_l > G_g$. The two phases can coexist if $\mu_l = \mu_g$ or $G_l = G_g$. For calculations of μ of the vdW fluid, it is most convenient to calculate it from the Helmholtz free energy in equation (4.3), yielding

$$
\begin{aligned}
\mu(\tau, V, N) &= \left(\frac{\partial F}{\partial N}\right)_{\tau,V} \\
&= -\tau \ln\left[\frac{n_Q(V - Nb)}{N}\right] + \frac{Nb\tau}{V - Nb} - \frac{2Na}{V} \\
&= -\tau \ln\left[\frac{V - Nb}{N}\right] + \frac{Nb\tau}{V - Nb} - \frac{2Na}{V} - \tau \ln n_Q.
\end{aligned}
\tag{4.10}
$$

Reference

[1] Haynes W M (ed) 2013 *CRC Handbook of Chemistry and Physics* 94th edn (Boca Raton, FL: CRC Press)

David C Johnston

Chapter 5

van der Waals equation of state, reduced variables and laws of corresponding states

5.1 van der Waals equation of state and reduced variables

Equation (4.8) can be written

$$\left(p + \frac{N^2 a}{V^2}\right)(V - Nb) = N\tau. \tag{5.1}$$

This is the vdW equation of state, which reduces to the ideal gas equation of state (the ideal gas law) $pV = N\tau$ when the molecular interaction parameters a and b are zero.

Using equations (4.4*d*), one can write equation (5.1) as

$$\left[\frac{p}{p_c} + \frac{3}{(V/V_c)^2}\right]\left(\frac{V}{V_c} - \frac{1}{3}\right) = \frac{8\tau}{3\tau_c}. \tag{5.2}$$

Note that N has disappeared as a state variable from this equation. Following the notation in [1], we define the reduced variables

$$\hat{p} \equiv \frac{p}{p_c} \qquad \hat{V} \equiv \frac{V}{V_c} \qquad \hat{\tau} \equiv \frac{\tau}{\tau_c} = \frac{T}{T_c}. \tag{5.3}$$

Then equation (5.2) becomes

$$\left(\hat{p} + \frac{3}{\hat{V}^2}\right)\left(3\hat{V} - 1\right) = 8\hat{\tau}, \tag{5.4}$$

which is the vdW equation of state written in reduced variables.

doi:10.1088/978-1-627-05532-1ch5

5.2 Laws of corresponding states

When two fluids are in 'corresponding states', they have the same set of three reduced parameters $\hat{p}$, $\hat{V}$ and $\hat{\tau}$. The differences between p_c, V_c and τ_c of different fluids are subsumed into the reduced parameters $\hat{p}$, $\hat{V}$ and $\hat{\tau}$. Therefore equation (5.4) is an example of a 'law of corresponding states' which is obeyed by all vdW fluids. Many other laws of corresponding states are derived below for the vdW fluid. From equation (5.4), the pressure versus volume and temperature is expressed in reduced variables as

$$\hat{p} = \frac{8\hat{\tau}}{3\hat{V} - 1} - \frac{3}{\hat{V}^2}. \tag{5.5}$$

Thus with decreasing $\hat{V}$, $\hat{p}$ diverges at $\hat{V} = 1/3$, which is the reduced volume at which the entire volume occupied by the fluid is filled with the hard-core molecules with no free volume remaining.

5.3 Pressure versus volume and versus density isotherms

Using equation (5.5), $\hat{p}$ versus $\hat{V}$ isotherms at several temperatures $\hat{\tau}$ are plotted in figure 5.1. One notices that for $\hat{\tau} > 1$ ($T > T_c$), the pressure monotonically decreases with increasing volume. This temperature region corresponds to a 'fluid' region where gas and liquid cannot be distinguished. At $T < T_c$ the isotherms show mechanically unstable behaviors in which the pressure increases with increasing volume over a certain range of p and V. This unstable region forms part of the volume region where liquid and gas coexist in equilibrium as further discussed below.

The order parameter for the liquid–gas phase transition is the difference in the number density $n = N/V$ between the liquid and gas phases [2, 3]. Using the third of equations (4.4d), one has

$$n = \frac{N}{V} = \frac{8p_c V_c}{3\tau_c V} = \frac{8p_c}{3\tau_c \hat{V}}, \tag{5.6a}$$

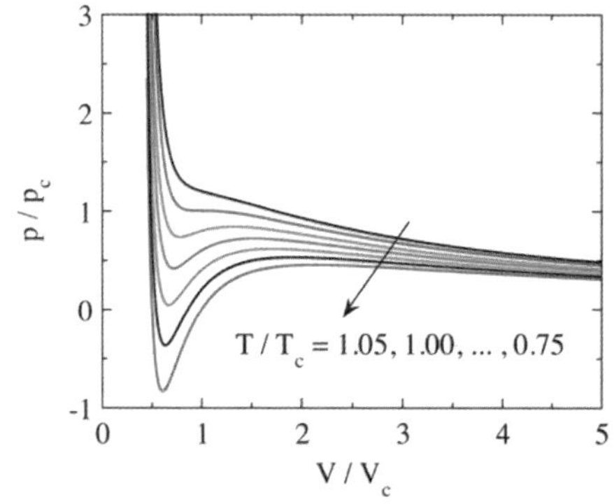

Figure 5.1. Reduced pressure $\hat{p} \equiv p/p_c$ versus reduced volume $\hat{V} \equiv V/V_c$ at several values of reduced temperature $\hat{\tau} \equiv \tau/\tau_c = T/T_c$ according to equation (5.5). The region on the far left corresponds to the liquid phase and the region to the far right corresponds to the gas or fluid phase depending on the temperature. The regions of negative pressure and positive $\mathrm{d}\hat{p}/\mathrm{d}\hat{V}$ are unphysical and correspond to regions of coexistence of the gas and liquid phases that are not taken into account in this figure.

where the second of equations (5.3) was used to obtain the last equality. The value of n_c at the critical point is obtained by setting $\hat{V} = 1$, yielding

$$n_c = \frac{8p_c}{3\tau_c}. \qquad (5.6b)$$

The critical concentration is also defined as

$$n_c = \frac{N}{V_c}. \qquad (5.7)$$

The reduced form of the number density analogous to those in equations (5.3) is obtained from equation (5.7) and the definitions $n = N/V$ and $\hat{V} = V/V_c$ as

$$\hat{n} \equiv \frac{n}{n_c} = \frac{1}{\hat{V}}. \qquad (5.8)$$

Using this expression, one can write equation (5.5) in terms of the reduced fluid number density as

$$\hat{p} = \frac{8\hat{\tau}\hat{n}}{3 - \hat{n}} - 3\hat{n}^2, \qquad (5.9)$$

with the restriction $\hat{n} < 3$ due to the excluded volume of the fluid.

Pressure versus temperature plots at constant volume and number density from equations (5.5) and (5.9) are linear as shown in figures 5.2(*a*) and 5.2(*b*), respectively. Isotherms of $\hat{p}$ versus $\hat{n}$ are shown in figure 5.3. The mechanically unstable and unphysical regions where $d\hat{p}/d\hat{n} < 0$ and $\hat{p} < 0$, respectively, correspond to similar regions in figure 5.1.

Volume versus temperature isobars obtained using equation (5.5) are shown in figure 5.4, where we first calculated $\hat{\tau}$ versus $\hat{V}$ and then inverted the axes. Some of these plots show mechanically unstable regions as in figures 5.1 and 5.3 that are associated with coexisting gas and liquid phases as discussed in chapter 6 below.

5.4 Influence of the van der Waals interactions on the pressure of the gas phase

There has been much discussion in the literature about whether the interactions between the molecules in the vdW gas (or undifferentiated fluid) phase increase the pressure or decrease the pressure of the gas compared to that of a (noninteracting) ideal gas at the same temperature and volume. For example Zemansky and Dittman [4], Berberan-Santos *et al* [5] and Stanley [2] state that the pressure decreases below that of the ideal gas due to the attactive interaction *a*, whereas Kittel and Kroemer [1] claim that the pressure increases. Tuttle has reviewed the history of this controversy, including a quote from van der Waals himself who evidently claimed that the pressure decreases [6]. Implicit in these statements is that the temperature and volume of the gas are not relevant to the argument. Here we show quantitatively that the vdW interactions *a* and *b* can either increase or decrease the pressure of a vdW gas compared to the ideal gas, depending on the temperature and volume of the gas.

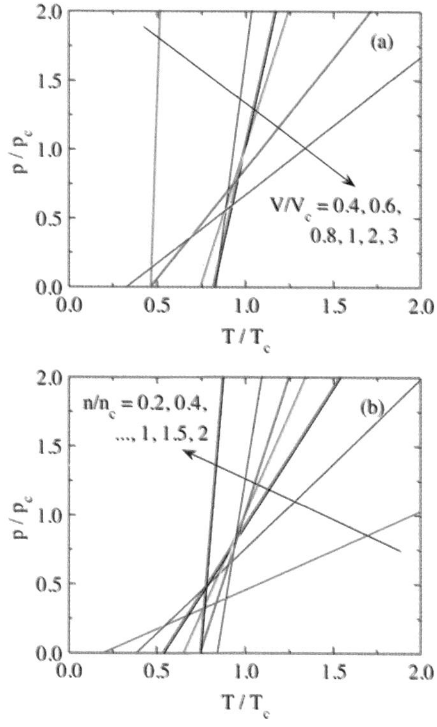

Figure 5.2. Reduced pressure $\hat{p} \equiv p/p_c$ versus reduced temperature $\hat{\tau} = T/T_c$ of the vdW fluid at (*a*) fixed reduced volume $\hat{V} \equiv V/V_c$ and (*b*) fixed reduced number density $\hat{n} \equiv n/n_c$ from equations (5.5) and (5.9), respectively.

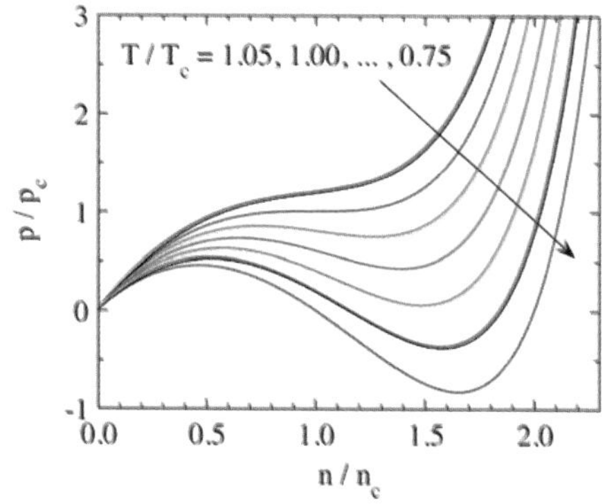

Figure 5.3. Reduced pressure $\hat{p} \equiv p/p_c$ versus reduced number density $\hat{n} \equiv n/n_c$ at several values of reduced temperature $\hat{\tau}$ according to equation (5.9). The region on the far left corresponds to the gas or fluid phase depending on the temperature and the region to the far right corresponds to the liquid phase. The regions of negative pressure and negative $\mathrm{d}\hat{p}/\mathrm{d}\hat{n}$ are unphysical for a homogeneous fluid and correspond to regions of coexisting gas and liquid phases.

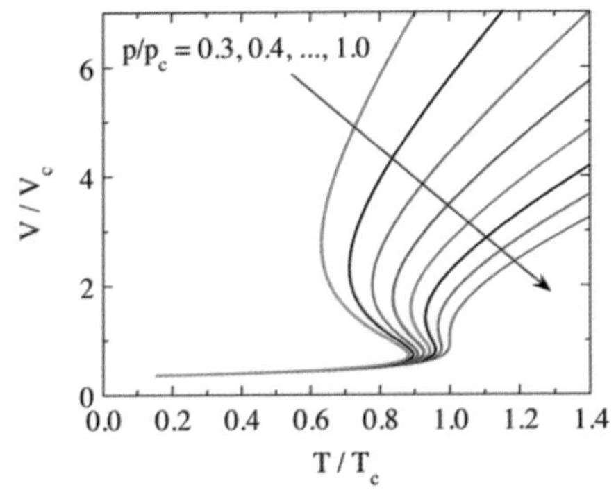

Figure 5.4. Reduced volume $\hat{V} = V/V_c$ versus reduced temperature $\hat{\tau} = T/T_c$ at several values of reduced pressure $\hat{p} = p/p_c$ according to equation (5.5). The regions of negative $\mathrm{d}\hat{V}/\mathrm{d}\hat{\tau}$ are unphysical and correspond to regions of coexisting gas and liquid phases.

The compression factor Z of a gas is defined as in equation (4.4*c*) to be

$$Z \equiv \frac{pV}{N\tau}. \tag{5.10}$$

For the ideal gas one has $Z_{\mathrm{IG}} = 1$. Using equation (4.8), the compression factor of the vdW gas is

$$Z = \frac{1}{1 - Nb/V} - \frac{Na}{V\tau}. \tag{5.11}$$

The deviation of Z from Z_{IG} is then

$$Z - 1 = \frac{Nb/V}{1 - Nb/V} - \frac{Na}{V\tau}. \tag{5.12}$$

In the current discussion the temperature and volume are constant as the vdW interactions are turned on and the right-hand side of equation (5.12) becomes nonzero. One sees from equation (5.12) that increasing b increases the pressure and increasing a decreases the pressure, where the ratio a/b is a fixed value for a given vdW gas according to equation (3.12). Therefore a competition occurs between these two effects on the pressure as the interactions are turned on. In reduced variables, equations (4.4*d*) give

$$\frac{Nb}{V} = \frac{1}{3\hat{V}} \qquad \frac{Na}{V\tau} = \frac{9}{8\hat{V}\hat{\tau}}. \tag{5.13}$$

Inserting these expressions into equation (5.12) gives

$$Z - 1 = \frac{\frac{1}{3\hat{V}}}{1 - \frac{1}{3\hat{V}}} - \frac{9}{8\hat{V}\hat{\tau}}. \tag{5.14}$$

One has the limits $0 < 1/\hat{V} < 3$. We recall that the first term on the right-hand side of equation (5.14) arises from the parameter b and the second one from a, rewritten in terms of reduced variables. The right-hand side is zero if $a = b = 0$.

Shown in figure 5.5(*a*) are isotherms of $Z - 1$ versus $1/\hat{V}$ plotted using equation (5.14) at several values of $1/\hat{\tau}$ as indicated. Expanded isotherms at low values of $1/\hat{V}$ and $1/\hat{\tau}$ are shown in figure 5.5(*b*). One sees from figure 5.5 that if $\hat{\tau}$ is below a certain value $\hat{\tau}_{\max}$, the pressure of the vdW gas is smaller than that of the ideal gas for a range of inverse volumes, whereas for $\hat{\tau} > \hat{\tau}_{\max}$ the pressure of the vdW gas increases with $1/\hat{V}$ above that of the ideal gas irrespective of the value of $\hat{\tau}$. Thus there is a crossover at $\hat{\tau}_{\max}$ where the initial slope $\partial(Z - 1)/\partial(1/\hat{V})$ goes from positive to negative with decreasing values of $\hat{\tau}$. By solving $\partial(Z - 1)/\partial(1/\hat{V}) = 0$ using equation (5.14), the crossover occurs at

$$\hat{\tau}_{\max} = \frac{27}{8}, \tag{5.15}$$

as indicated by the inverse of $\hat{\tau}_{\max}$ in figure 5.5.

To summarize, if the temperature of a vdW gas is less than $\hat{\tau}_{\max}$, there is a range of inverse volumes (or number densities) over which the molecular interactions cause the pressure to be less than that of the ideal gas, whereas if the temperature is greater than

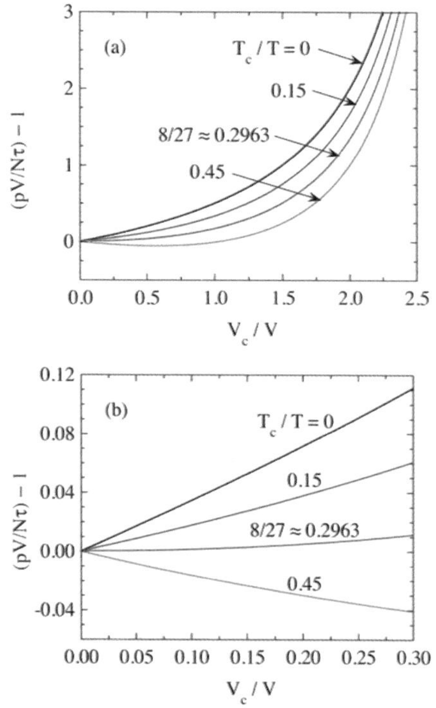

Figure 5.5. (*a*) Isotherms of the quantity $Z - 1 = (pV/N\tau) - 1$ versus inverse reduced volume $1/\hat{V} = V_c/V$ at the inverse temperatures $1/\hat{\tau} = T_c/T$ indicated. (*b*) Expanded plots of $Z - 1$ versus $1/\hat{V}$ at small values of $1/\hat{V}$ from (*a*).

$\hat{\tau}_{\max}$, the interactions increase the pressure above that of the ideal gas irrespective of the value of the inverse volume. To find this maximum value of the inverse volume versus $\hat{\tau}$, one can set $Z - 1 = 0$ in equation (5.14) and solve for $(1/\hat{V})_{\max}$, yielding

$$\left(\frac{1}{\hat{V}}\right)_{\max} = 3 - \frac{8\hat{\tau}}{9} \quad (\hat{\tau} < \hat{\tau}_{\max}). \tag{5.16}$$

If $\hat{\tau} < 1$, the possibility of liquifaction of the gas exists as discussed below, so the discussion here refers only to the gas phase in this temperature range.

We conclude that the vdW interaction parameters can give rise to either an increase or a decrease in the pressure of a vdW gas relative to that of an ideal gas at the same volume and temperature, depending on the actual values of the volume and temperature.

5.5 Boyle temperature

From equation (5.14), the temperature $\hat{\tau}_{\rm B}$ at which $Z - 1 = 0$, at which the net effect of the molecular interactions on the compression factor compared to that of the ideal gas is zero, is

$$\hat{\tau}_{\rm B} = \frac{9}{8}\left(3 - \frac{1}{\hat{V}}\right), \tag{5.17}$$

where $\hat{\tau}_{\rm B}$ is known as the Boyle temperature. At large volumes the Boyle temperature approaches the limit $\hat{\tau}_{\rm B\,max} = \hat{\tau}_{\max} = 27/8$ in equation (5.15) and it decreases monotonically from there with decreasing volume. For the minimum value of $\hat{V}$ of 1/3 (at which the free volume goes to zero), equation (5.17) gives the minimum value of the Boyle temperature as $\hat{\tau}_{\rm B\,min} = 0$.

It is sometimes stated that the Boyle temperature is the temperature at which a gas with molecular interactions behaves like an ideal gas. This definition is misleading, because it only applies to the compression factor and not to thermodynamic properties like the heat capacity at constant pressure $C_{\rm p}$, the isothermal compressibility $\kappa_{\rm T}$ or the coefficient of volume expansion α. We show in the following section 5.6 that the molecular interactions have nonzero influences on these thermodynamic properties at all finite temperatures and volumes of the vdW fluid.

5.6 Thermodynamic properties expressed in reduced variables

5.6.1 Internal energy

One can write the internal energy in equation (4.6) in terms of the reduced variables in equations (4.4) and also in terms of $\hat{n} = 1/\hat{V}$ defined in (5.8) as

$$\begin{aligned}\frac{U}{p_{\rm c}V_{\rm c}} &= 4\hat{\tau} - \frac{3}{\hat{V}}\\ &= 4\hat{\tau} - 3\hat{n}.\end{aligned} \tag{5.18}$$

At the critical point, one obtains

$$\frac{U_c}{p_c V_c} = 1 \quad \left(\hat{\tau} = \hat{V} = \hat{n} = \hat{p} = 1\right). \tag{5.19}$$

5.6.2 Helmholtz free energy

To write F in terms of reduced variables, we first write the quantum concentration in equation (2.2b) as

$$n_Q = n_{Qc}\,\hat{\tau}^{3/2}, \tag{5.20}$$

where

$$n_{Qc} \equiv \left(\frac{m\tau_c}{2\pi\hbar^2}\right)^{3/2}. \tag{5.21}$$

Then in terms of the reduced variables, the Helmholtz free energy in equation (4.3) becomes

$$\frac{F}{p_c V_c} = -\frac{8\hat{\tau}}{3}\left\{\ln\left[x_c\,\hat{\tau}^{3/2}\left(3\hat{V} - 1\right)\right] + 1\right\} - \frac{3}{\hat{V}}, \tag{5.22a}$$

where the dimensionless variable x_c is

$$x_c \equiv \frac{n_{Qc}\tau_c}{8p_c}. \tag{5.22b}$$

5.6.3 Entropy and enthalpy

The entropy in equation (4.5) becomes

$$\begin{aligned}\frac{S}{Nk_B} &= \ln\left[x_c\,\hat{\tau}^{3/2}\left(3\hat{V} - 1\right)\right] + \frac{5}{2}\\ &= \ln\left[x_c\,\hat{\tau}^{3/2}\left(3 - \hat{n}\right)/\hat{n}\right] + \frac{5}{2}.\end{aligned} \tag{5.23}$$

The entropy diverges to $-\infty$ at $\hat{\tau} \to 0$, which violates the third law of thermodynamics and thus shows that the vdW fluid is in the classical regime just as the ideal gas is.

The enthalpy in equation (4.9) becomes

$$\begin{aligned}\frac{H}{p_c V_c} &= \frac{4\hat{\tau}\left(5\hat{V} - 1\right)}{3\hat{V} - 1} - \frac{6}{\hat{V}}\\ &= \frac{4\hat{\tau}\left(5 - \hat{n}\right)}{3 - \hat{n}} - 6\hat{n}.\end{aligned} \tag{5.24}$$

At the critical point $\hat{\tau} = \hat{p} = \hat{V} = \hat{n} = 1$, the reduced critical enthalpy $H_c/(p_c V_c)$ is given by equation (5.24) as

$$\frac{H_c}{p_c V_c} = 2 \quad \text{(at the critical point)}. \tag{5.25}$$

Equations (5.18) and (5.24) are laws of corresponding states. However equations (5.22*a*) and (5.23) are not because they explicitly depend on the mass m of the molecules in the particular fluid considered. On the other hand, the difference in entropy per particle $\Delta S/(Nk_B)$ between one reduced state of a vdW fluid and another is a law of corresponding states. Taking the reference state to be the critical point at which $\hat{\tau}^{3/2}(3\hat{V} - 1) = 2$, equation (5.23) yields

$$\begin{aligned}\frac{\Delta S}{Nk_B} &\equiv \frac{S(\hat{\tau}, \hat{V}) - S(1, 1)}{Nk_B} \\ &= \ln\left[\hat{\tau}^{3/2}(3\hat{V} - 1)/2\right] \\ &= \ln\left[\hat{\tau}^{3/2}(3 - \hat{n})/(2\hat{n})\right]. \end{aligned} \tag{5.26}$$

5.6.4 Isothermal compressibility

The isothermal compressibility κ_T is given by equation (2.7). In the reduced units in equations (5.3) one obtains

$$\frac{1}{\kappa_T p_c} = -\hat{V}\left(\frac{\partial \hat{p}}{\partial \hat{V}}\right)_{\hat{\tau}}. \tag{5.27a}$$

One can write the partial derivative on the right-hand side as [2]

$$\left(\frac{\partial \hat{p}}{\partial \hat{V}}\right)_{\hat{\tau}} = -\left(\frac{\partial \hat{p}}{\partial \hat{\tau}}\right)_{\hat{V}}\left(\frac{\partial \hat{\tau}}{\partial \hat{V}}\right)_{\hat{p}}, \tag{5.27b}$$

so equation (5.27*a*) can also be written

$$\frac{1}{\kappa_T p_c} = \hat{V}\left(\frac{\partial \hat{p}}{\partial \hat{\tau}}\right)_{\hat{V}}\left(\frac{\partial \hat{\tau}}{\partial \hat{V}}\right)_{\hat{p}}. \tag{5.27c}$$

Utilizing the expression for the reduced pressure for the vdW fluid in equation (5.5), equation (5.27*a*) gives

$$\kappa_T p_c = \frac{(3\hat{V} - 1)^2 \hat{V}^2/6}{4\hat{\tau}\hat{V}^3 - (3\hat{V} - 1)^2}. \tag{5.28a}$$

In terms of $\hat{n} = 1/\hat{V}$, equation (5.28*a*) becomes

$$\kappa_T p_c = \frac{(3 - \hat{n})^2/(6\hat{n})}{4\hat{\tau} - \hat{n}(3 - \hat{n})^2}. \tag{5.28b}$$

Using equation (4.4*c*), a Taylor series expansion of equation (5.28*a*) about $1/\hat{V} = 0$ in powers of $1/\hat{V}$ gives

$$\kappa_T = \frac{3\hat{V}}{8\hat{\tau}p_c}\left[1 + \frac{27 - 8\hat{\tau}}{12\hat{\tau}\hat{V}} + O\left(\frac{1}{\hat{V}^2}\right)\right], \tag{5.28c}$$

where the prefactor is

$$\frac{3\hat{V}}{8\hat{\tau}p_c} = \frac{V}{Nk_B T}, \tag{5.28d}$$

which is the result for the ideal gas in equation (2.7) that $\kappa_T = 1/p = V/(Nk_B T)$. Thus in the limit $\hat{V} \to \infty$ one obtains the expression for the ideal gas.

5.6.5 Volume thermal expansion coefficient

The volume thermal expansion coefficient α is defined in equation (2.8). In reduced units one has

$$\frac{\alpha\tau_c}{k_B} = \frac{1}{\hat{V}}\left(\frac{\partial \hat{V}}{\partial \hat{\tau}}\right)_{\hat{p}}. \tag{5.29}$$

Comparing equations (5.27*c*) and (5.29) shows that

$$\frac{\alpha\tau_c/k_B}{\kappa_T p_c} = \left(\frac{\partial \hat{p}}{\partial \hat{\tau}}\right)_{\hat{V}}. \tag{5.30}$$

Utilizing the expression for the reduced pressure of the vdW fluid in equation (5.5), equation (5.29) gives

$$\frac{\alpha\tau_c}{k_B} = \frac{4\left(3\hat{V} - 1\right)\hat{V}^2/3}{4\hat{\tau}\hat{V}^3 - \left(3\hat{V} - 1\right)^2}. \tag{5.31a}$$

In terms of the reduced number density $\hat{n} = 1/\hat{V}$ one obtains

$$\frac{\alpha\tau_c}{k_B} = \frac{4\left(3 - \hat{n}\right)/3}{4\hat{\tau} - \hat{n}\left(3 - \hat{n}\right)^2}. \tag{5.31b}$$

A Taylor series expansion of equation (5.31*a*) about $1/\hat{V} = 0$ in powers of $1/\hat{V}$ gives

$$\frac{\alpha\tau_c}{k_B} = \frac{1}{\hat{\tau}}\left[1 + \frac{27 - 4\hat{\tau}}{12\hat{\tau}\hat{V}} + O\left(\frac{1}{\hat{V}^2}\right)\right]. \tag{5.31c}$$

In the limit of large volumes or small concentrations one obtains

$$\frac{\alpha\tau_c}{k_B} = \frac{1}{\hat{\tau}} \quad \left(\hat{V} \to \infty\right), \tag{5.31d}$$

which agrees with the ideal gas value for the thermal expansion coefficient in equation (2.8).

Comparing equations (5.31*a*) and (5.28*a*) shows that the dimensionless reduced values of κ_T and α are simply related according to

$$\frac{\alpha\tau_c/k_B}{\kappa_T p_c} = \frac{8}{3\hat{V} - 1} = \frac{8\hat{n}}{3 - \hat{n}}, \tag{5.32}$$

which using the expression (5.5) for the pressure is seen to be in agreement with the general equation (5.30).

5.6.6 Heat capacity at constant pressure

The heat capacity at constant pressure C_p and at constant volume C_V are related according to the thermodynamic relation in equation (2.9). In reduced units, this equation becomes

$$C_p - C_V = k_B\left(\frac{p_c V_c}{\tau_c}\right)\frac{\hat{\tau}\hat{V}\left(\alpha\tau_c/k_B\right)^2}{\left(\kappa_T p_c\right)}. \tag{5.33a}$$

Using equation (4.4*c*) one then obtains

$$\frac{C_p - C_V}{Nk_B} = \frac{3\hat{\tau}\hat{V}\left(\alpha\tau_c/k_B\right)^2}{8\left(\kappa_T p_c\right)}. \tag{5.33b}$$

The expression for C_V in equation (4.7) for a single-phase vdW fluid then gives

$$\frac{C_p}{Nk_B} = \frac{3}{2} + \frac{3\hat{\tau}\hat{V}\left(\alpha\tau_c/k_B\right)^2}{8\left(\kappa_T p_c\right)}. \tag{5.33c}$$

Utilizing equations (5.31*a*) and (5.32), the heat capacity at constant pressure in equation (5.33*c*) for the vdW fluid simplifies to

$$\frac{C_p}{Nk_B} = \frac{3}{2} + \frac{1}{1 - \dfrac{\left(3\hat{V} - 1\right)^2}{4\hat{\tau}\hat{V}^3}}. \tag{5.34a}$$

The C_p can be written in terms of the reduced number density $\hat{n} = 1/\hat{V}$ as

$$\frac{C_p}{Nk_B} = \frac{3}{2} + \frac{4\hat{\tau}}{4\hat{\tau} - \hat{n}\left(3 - \hat{n}\right)^2}. \tag{5.34b}$$

A Taylor series expansion of equation (5.34*a*) about $1/\hat{V} = 0$ in powers of $1/\hat{V}$ gives

$$\frac{C_p}{Nk_B} = \frac{5}{2} + \frac{9}{4\hat{\tau}\hat{V}} + O\left(\frac{1}{\hat{V}^2}\right). \tag{5.34c}$$

In the limit $\hat{V} \to \infty$, this equation gives the ideal gas expression for C_p in equation (2.9).

As one approaches the critical point with $\hat{p} \to 1, \hat{V} \to 1, \hat{n} \to 1$ and $\hat{\tau} \to 1$, one obtains κ_T, α, $C_p \to \infty$. These critical behaviors are quantitatively discussed in chapter 8 below.

The latent heat (enthalpy) of vaporization L is defined as

$$L = T\Delta S_X, \tag{5.35}$$

where ΔS_X is the change in entropy of the system when liquid is completely converted to gas at constant temperature. Using equations (4.4c) and (5.3), L can be written in dimensionless reduced form as

$$\frac{L}{p_c V_c} = \left(\frac{8\hat{\tau}}{3}\right)\frac{\Delta S_X}{Nk_B}. \tag{5.36}$$

5.6.7 Chemical potential

Using equations (4.4d) and (5.3), one can express $\mu(\tau, V, N)$ in equation (4.10) in reduced variables as

$$\frac{\mu}{\tau_c} = -\hat{\tau}\ln\left(3\hat{V} - 1\right) + \frac{\hat{\tau}}{3\hat{V} - 1} - \frac{9}{4\hat{V}} - \hat{\tau}\ln X, \tag{5.37a}$$

$$X \equiv \frac{\tau_c n_Q}{8p_c}. \tag{5.37b}$$

In terms of the reduced number density $\hat{n} = 1/\hat{V}$, equation (5.37a) becomes

$$\frac{\mu}{\tau_c} = -\hat{\tau}\ln\left(\frac{3 - \hat{n}}{\hat{n}}\right) + \frac{\hat{\tau}\hat{n}}{3 - \hat{n}} - \frac{9\hat{n}}{4} - \hat{\tau}\ln X. \tag{5.37c}$$

The above terms in X depend on the particular vdW fluid being considered.

We add and subtract $\ln(2e^{-1/2})$ from the right-hand side of equation (5.37a), yielding

$$\frac{\mu}{\tau_c} = -\hat{\tau}\ln\left(\frac{3\hat{V} - 1}{2e^{-1/2}}\right) + \frac{\hat{\tau}}{3\hat{V} - 1} - \frac{9}{4\hat{V}}$$
$$- \hat{\tau}\ln\left(2e^{-1/2}X\right). \tag{5.38}$$

For processes at constant $\hat{\tau}$, the last (gas-dependent) term in equation (5.38) just has the effect of shifting the origin of μ/τ_c as $\hat{\tau}$ is changed. When plotting μ/τ_c versus $\hat{p}$ or μ/τ_c versus $\hat{V}$ isotherms, we set that constant to zero, yielding

$$\frac{\mu}{\tau_c} = -\hat{\tau}\ln\left(\frac{3\hat{V} - 1}{2e^{-1/2}}\right) + \frac{\hat{\tau}}{3\hat{V} - 1} - \frac{9}{4\hat{V}}. \tag{5.39}$$

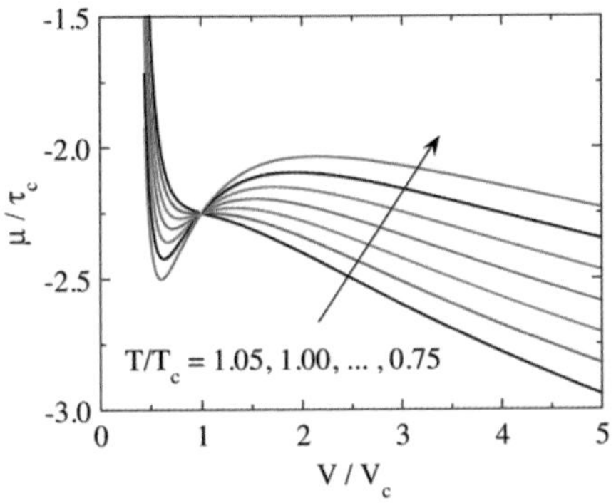

Figure 5.6. Reduced chemical potential μ/τ_c versus reduced volume $\hat{V} \equiv V/V_c$ at several values of reduced temperature $\hat{\tau} \equiv \tau/\tau_c$ according to equation (5.39). For $\hat{\tau} < 1$, the region on the far left corresponds to the liquid phase and the region to the far right corresponds to the gas phase, with a region of coexistence between them.

Including the factor of $2e^{-1/2}$ in equation (5.39) causes the μ/τ_c versus $\hat{V}$ isotherms to all cross at $\hat{V} = 1$. Isotherms of μ/τ_c versus $\hat{V}$ obtained using equation (5.39) are plotted in figure 5.6 for the critical temperature and several adjacent temperatures.

References

[1] Kittel C and Kroemer H 1980 *Thermal Physics* 2nd edn (New York: Freeman)

[2] Stanley H E 1971 *Introduction to Phase Transitions and Critical Phenomena* (New York: Oxford Science)

[3] Kadanoff L P, Götze W, Hamblen D, Hecht R, Lewis E A S, Palciauskas V V, Rayl M and Swift J 1967 Static phenomena near critical points: theory and experiment *Rev. Mod. Phys.* **39** 395

[4] Zemansky M W and Dittman R H 1997 *Heat and Thermodynamics* 7th edn (New York: McGraw-Hill)

[5] Berberan-Santos M N, Bodunov E N and Pogliani L 2008 The van der Waals equation: analytical and approximate solutions *J. Math. Chem.* **43** 1437

[6] Tuttle E R 1975 The cohesion term in van der Waals's equation of state *Am. J. Phys.* **43** 644

David C Johnston

Chapter 6

Equilibrium pressure–volume, temperature–volume and pressure–temperature phase diagrams

By combining the data in figures 5.1 and 5.6, one can plot μ versus p isotherms with V as an implicit parameter, as shown in figure 6.1. Since here $\mu = \mu(\tau, p)$ as in equation (2.12a), in equilibrium the state occurs with the lowest Gibbs free energy and therefore also the lowest chemical potential.

Following Reif [1], certain points on a p–V isotherm at $\hat{\tau} < 1$ are shown in the top panel of figure 6.2 and compared with the corresponding points on a plot of μ versus p in the bottom panel. Starting from the bottom left of the bottom panel of figure 6.2, at low pressure the stable phase is seen to be the gas phase. As the pressure increases, a region occurs at which the chemical potential of the gas and liquid become the same, at the point X at pressure p_X, which signals entry into a triangle-shaped unstable region of the plot which the system does not enter in thermal equilibrium. The pressure p_X is a constant pressure part of the p–V isotherm at which the gas and liquid coexist as indicated by the horizontal line in the top panel. The system remains at constant pressure at the point X in the bottom panel as the system volume decreases until all the gas is converted to liquid. At higher pressures, the pure liquid has the lower chemical potential as indicated in the bottom panel.

Essential variables in calculations of the thermodynamic properties of the vdW fluid are the reduced equilibrium pressure $\hat{p}_X$ for coexistence of gas and liquid phases at a given $\hat{\tau}$ and the associated reduced volumes $\hat{V}_C$, $\hat{V}_D$, $\hat{V}_F$, and $\hat{V}_G$ in figure 6.2. Using these values and the equations for the thermodynamic variables and properties, the equilibrium and nonequilibrium properties versus temperature, volume or pressure can be calculated and the various phase diagrams constructed. The condition for the coexistence of the liquid and gas phases is that their chemical

doi:10.1088/978-1-627-05532-1ch6

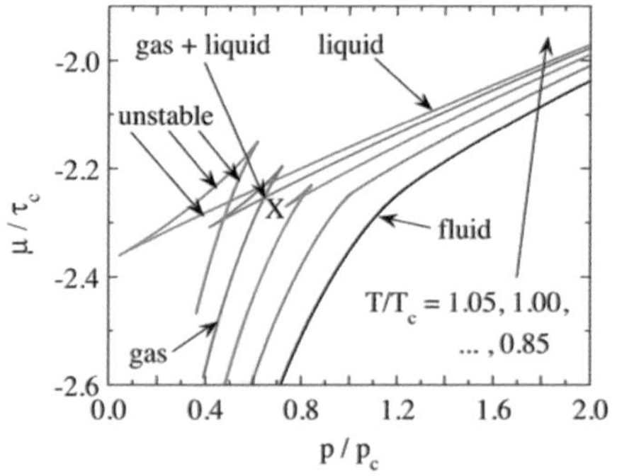

Figure 6.1. Reduced chemical potential μ/τ_c versus reduced pressure $\hat{p} \equiv p/p_c$, with the volume of the system as an implicit parameter, at several values of reduced temperature $\hat{\tau} \equiv \tau/\tau_c = T/T_c$. The low-pressure and high-volume pure gas region is at the lower left and the high-pressure and low-volume pure liquid region is at the upper right. As the volume decreases on moving upwards along an isotherm, the pressure increases until the system encounters the intersection with an unstable or metastable triangle-shaped part of the isotherm, labeled as the point X. At this point the chemical potentials of the gas and liquid are the same. As the volume decreases further, if the system is in equilibrium the pressure remains constant at this point until all the gas is converted to liquid. Then the pressure increases again as the volume of liquid decreases. The fluid phase, where liquid and gas cannot be distinguished, occurs at temperatures T at and above the critical temperature T_c.

potentials $\mu(\hat{\tau}, \hat{p})$, temperatures and pressures must be the same at their respective volumes $\hat{V}_G$ and $\hat{V}_C$ in figure 6.2, where $\mu(\hat{\tau}, \hat{V})/\tau_c$ is given in equation (5.39). This requirement allows $\hat{p}_X$ to be determined.

To determine the values of $\hat{p}_X$, $\hat{V}_C$, $\hat{V}_D$, $\hat{V}_F$ and $\hat{V}_G$ in figure 6.2, where gas and liquid phases coexist in equilibrium, one can use a parametric solution in which $\hat{p}$ and μ/τ_c are calculated at fixed $\hat{\tau}$ using $\hat{V}$ as an implicit parameter and thereby express μ/τ_c versus $\hat{p}$ at fixed $\hat{\tau}$. From the numerical $\mu(\hat{\tau}, \hat{p})$ data, one can then determine the values of the above four reduced volumes and then the value of p_X from $\hat{V}_C$ or $\hat{V}_G$ and the vdW equation of state. The following steps are carried out for each specified value of $\hat{\tau}$ to implement this sequence of calculations.

1. The two volumes $\hat{V}_D$ and $\hat{V}_F$ at the maximum and minimum, respectively, of the S-shaped region of the p–V plot in figure 6.2 are determined by solving equation (5.5) for the two volumes at which $\partial\hat{p}/\partial\hat{V} = 0$. These two volumes enclose the mechanically unstable region of phase separation of the gas and liquid phases since the isothermal compressibility in equation (2.7) is negative in this region.
2. The pressure $\hat{p}_2$ at the volume $\hat{V}_D$ is determined from the equation of state (5.5).
3. The volume $\hat{V}_H$ is determined by solving equation (5.5) for the two volumes at pressure $\hat{p}_2$ (the other one, $\hat{V}_D$, is already calculated in step 1). These two volumes are needed to set the starting values of the numerical calculations of the volumes $\hat{V}_C$ and $\hat{V}_G$ in the next step.

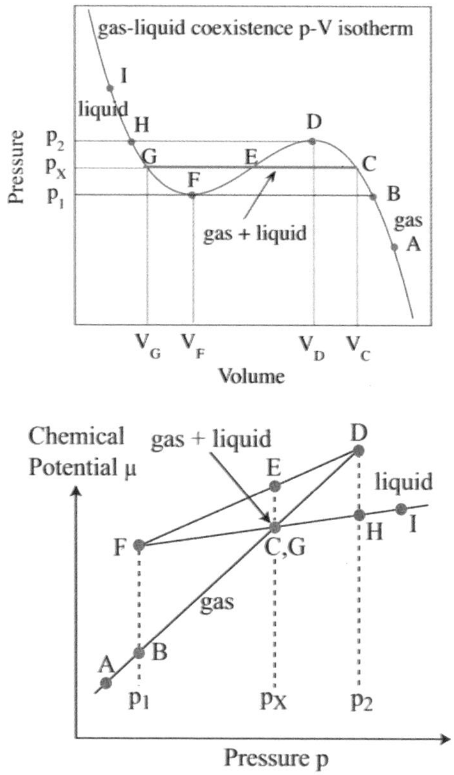

Figure 6.2. (Top panel) A schematic pressure versus volume (p–V) isotherm at $T < T_c$ such as in figure 5.1 showing coexistence between the gas and liquid phases as the horizontal line. Various points on the isotherm are labeled. With decreasing volume, in equilibrium the system follows the path A–B–C–E–G–H–I. One starts with pure gas at point A. Liquid starts to form at point C and all the gas is converted to liquid upon reaching point G. The system is pure liquid along the path G–H–I. (Bottom panel) A schematic chemical potential μ versus pressure isotherm such as shown in figure 6.1, with points labeled as in the top panel. Starting at low pressure (large volume), in equilibrium the system follows the path with the lowest chemical potential, i.e. A–B–(C,G)–H–I. The paths with decreasing volume C–D and F–G are metastable regions and the region D–E–F is mechanically unstable to phase separation. The temperatures of the isotherms in the two panels are the same.

4. The volumes $\hat{V}_C$ and $\hat{V}_G$ are determined by solving two simultaneous equations which equate the pressure and chemical potential of the gas and liquid phases at these two volumes at a fixed set temperature, respectively:

$$\hat{p}\left(\hat{\tau},\, \hat{V}_G\right) = \hat{p}\left(\hat{\tau},\, \hat{V}_C\right), \tag{6.1a}$$

$$\frac{\mu\left(\hat{\tau},\, \hat{V}_G\right)}{\tau_c} = \frac{\mu\left(\hat{\tau},\, \hat{V}_C\right)}{\tau_c}. \tag{6.1b}$$

The `FindRoot` utility of `Mathematica` is efficient in finding the solutions for $\hat{V}_G$ and $\hat{V}_C$ if appropriate starting values for these parameters are given.

The starting values we used were $0.97\hat{V}_H$ and $1.1\hat{V}_D$, respectively, where the volumes $\hat{V}_H$ and $\hat{V}_D$ are obtained from steps 3 and 1, respectively.

5. The pressure $\hat{p}_X$ at which the gas and liquid are in equilibrium at a given temperature is calculated from either $\hat{V}_C$ or $\hat{V}_G$ using the equation of state (5.5).

Representative values of the above reduced parameters calculated versus reduced temperature are given in table A.1 of appendix A.

6.1 Pressure–volume phase diagram and Maxwell construction

By solving for the pressure p_X versus temperature at which the gas and liquid phases coexist as described above, one can derive equilibrium pressure–volume isotherms. Representative isotherms are shown in figure 6.3 for $T/T_c = 1$ (critical temperature), 0.95, 0.90, 0.85 and 0.80. The locus of the end-points of the two-phase regions at constant pressure versus temperature is shown as the dashed curve in figure 6.3, which bounds a two-phase region highlighted in yellow above which the pure gas, pure liquid and undifferentiated fluid phases occur. Thus one obtains the pressure versus volume phase diagram shown in figure 6.4(*a*).

The pressure p_X at which the horizontal two-phase line occurs in the upper panel of figure 6.2 can be shown to satisfy the so-called Maxwell construction as follows. According to equation (2.12*b*) with constant τ and N, the differential of the Gibbs free energy is $dG = V\,dp$, so the difference in G between two pressures along a path in the V–p plane is

$$\Delta G = N\Delta\mu = G(p_2, V_2) - G(p_1, V_1) = \int_{p_1}^{p_2} V(p)\mathrm{d}p. \tag{6.2}$$

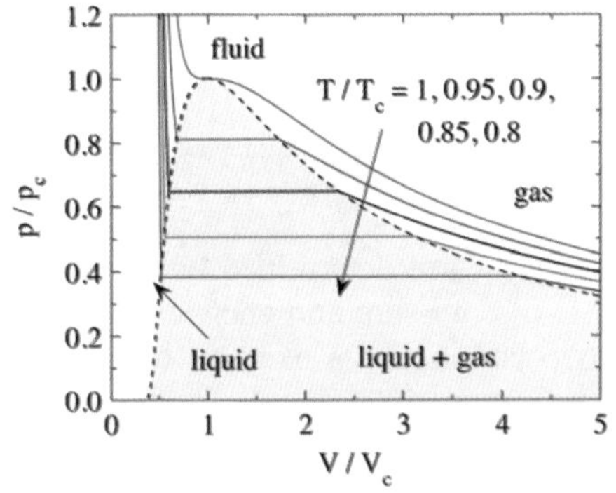

Figure 6.3. Equilibrium pressure versus volume isotherms showing the pure gas regime on the right, the liquid–gas coexistence region in the middle and the pure liquid regime on the left. The dashed curve is the boundary between the regions of pure liquid, liquid + gas and pure gas in the equilibrium p–V phase diagram. The yellow-shaded region bounded from above by the dashed curve is therefore the region of coexistence of the liquid and gas phases in the p–V phase diagram that is also shown in figure 6.4(*a*). For temperatures above the reduced critical temperature $T/T_c = 1$, there is no phase separation and no physical distinction between gas and liquid phases; this undifferentiated phase is denoted as the fluid phase.

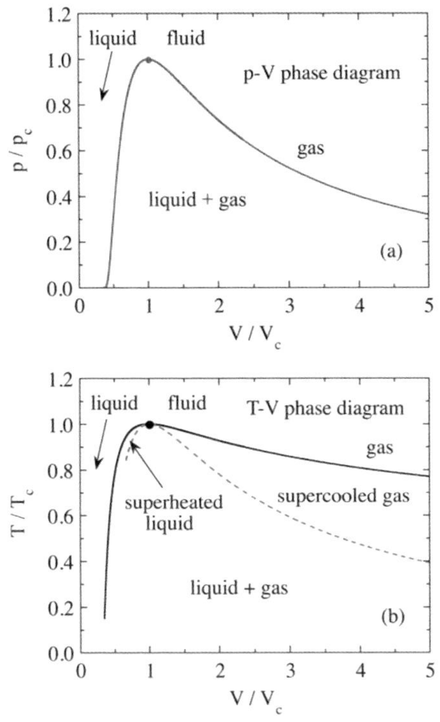

Figure 6.4. Equilibrium phase diagrams of the vdW fluid in the (*a*) pressure–volume (*p–V*) plane and (*b*) temperature–volume (*T–V*) plane. The critical point is denoted by a filled circle in each panel. In (*b*), the regions of metastable superheated liquid and supercooled gas, derived as in figure 9.1 below, are also shown with boundaries from the numerical data in table A.1 in appendix A that are defined by the solid black equilibrium curve and the respective colored dashed metastable curves as shown.

As shown in figure 6.5(*a*), this integral is the integral along the path from point C to point G. The part of the area beneath the curve from C to D that lies below the curve from D to E is canceled out because the latter area is negative. Similarly, the part of the negative area from E to F that lies below the path from F to G is canceled out by the positive area below the path from F to G. Therefore the net area from C to G is the sum of the positive hatched area to the right of the vertical line and the negative hatched area to the left of the vertical line that are shown in figure 6.5(*a*). Since the vertical line represents equilibrium between the gas and liquid phases, for which the chemical potentials and Gibbs free energies are the same, one has $\Delta G = 0$ and hence the algebraic sum of the two hatched areas is zero. This means that the magnitudes of the two hatched areas have to be the same.

Transferring this information to the corresponding *p–V* diagram in figure 6.5(*b*), one requires that the magnitudes of the same two hatched areas shown in that figure have to be equal. This is Maxwell's construction [2]. In terms of the numerical

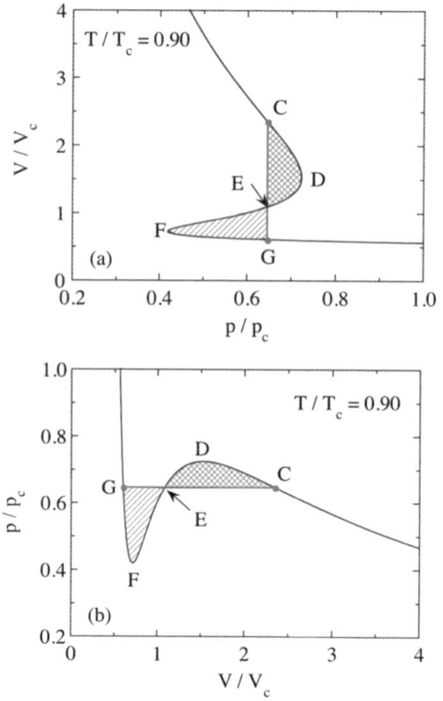

Figure 6.5. (a) Reduced volume $\hat{V} = V/V_c$ versus reduced pressure $\hat{p} = p/p_c$ at reduced temperature $\hat{\tau} = T/T_c = 0.9$, for which $\hat{V}_G = 0.603$ and $\hat{V}_C = 2.349$. The letter designations of points on the curve are the same as in figure 6.2. The vertical red line at reduced pressure $\hat{p}_X = 0.647$ is the equilibrium pressure of liquid–gas coexistence at $\hat{\tau} = 0.9$. According to equation (6.2), the net area under the curve from point C, where the fluid is pure gas, to G, where the fluid is pure liquid, is proportional to the change $\Delta\mu$ in the chemical potential between these two points, which is equal to zero for the region of liquid–gas coexistence from C to G. This $\Delta\mu$ is proportional to the net area of $\hat{V}$ versus $\hat{p}$ along the path from point C to point G, which in turn is the sum of the positive hatched area to the left of the path C–D–E, and the negative hatched area to the right of the path E–F–G. Therefore the magnitudes of the first and second areas must be the same. This requirement is drawn on the corresponding p–V diagram in (b), where the magnitudes of the two hatched areas above and below the horizontal line at $\hat{p}_X = 0.647$ must be the same. This is Maxwell's construction [2].

integral of a $\hat{p}$ versus $\hat{V}$ isotherm over the two-phase region at temperature $\hat{\tau}$, Maxwell's construction states that

$$\int_{\hat{V}_G}^{\hat{V}_C} \left[\hat{p}\left(\hat{\tau},\, \hat{V}\right) - \hat{p}_X(\hat{\tau})\right] \mathrm{d}\hat{V} = 0, \tag{6.3a}$$

or equivalently

$$\int_{\hat{V}_G}^{\hat{V}_C} \hat{p}\left(\hat{\tau},\, \hat{V}\right) \mathrm{d}\hat{V} = \hat{p}_X(\hat{\tau})\left(\hat{V}_C - \hat{V}_G\right). \tag{6.3b}$$

From equation (6.3*b*), the area under a $\hat{p}$ versus $\hat{V}$ curve as in figure 6.2 or figure 6.3 from $\hat{V}_G$ to $\hat{V}_C$ is equal to the area under the horizontal line between the same two volumes.

6.2 Volume–temperature phase diagram

Equilibrium volume versus temperature isobars are shown in figure 6.6 for reduced pressures $\hat{p} = 0.3, 0.4, \ldots, 1.0$. The isobars illustrate the first-order increase in volume at the liquid to gas transition temperature for each pressure with $\hat{\tau} < 1$. For $\hat{\tau} \geqslant 1$ only the undifferentiated fluid phase occurs.

The temperature versus volume phase diagram is shown in figure 6.4(*b*). Also included in figure 6.4(*b*) are regions in which metastable superheated liquid and supercooled gas occur, as discussed below in chapter 9. It may seem counterintuitive that both dashed lines lie *below* the equilibrium curve. However, in figure 9.1(*b*) below, it is shown that the superheated liquid has a larger volume without much change in temperature, resulting in the superheated metastable region in figure 6.4(*b*) being below the equilibrium curve.

6.3 Lever rule

Each $\hat{p}$ versus $\hat{V}$ isotherm at $T < T_c$ in figure 6.3 contains a horizontal line over a range of volumes in which the liquid and gas phases coexist. The left end of a line corresponds to pure liquid and the right end to pure gas. The same situation occurs for the vertical coexistence lines in the $\hat{V}$ versus $\hat{\tau}$ phase diagram in figure 6.6. One can determine the molar ratio of this mixture in terms of a given volume $\hat{V}_0$ in the coexistence region with $\hat{V}_l \leqslant \hat{V}_0 \leqslant \hat{V}_g$, where the symbols V_l, $\hat{V}_l$ and V_g, $\hat{V}_g$ are reserved for the actual and reduced volumes at the pure liquid and pure gas ends of the coexistence line, respectively.

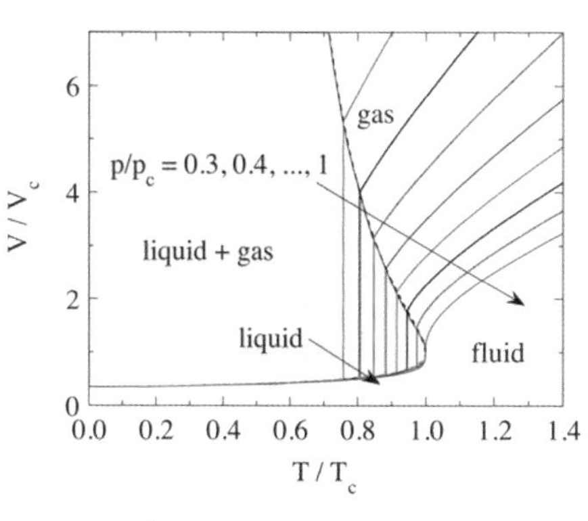

Figure 6.6. Equilibrium reduced volume $\hat{V} = V/V_c$ versus reduced temperature $\hat{\tau} = \tau/\tau_c = T/T_c$ isobars at the indicated reduced pressures $\hat{p} = p/p_c$. The dashed curve separates the pure liquid, liquid + gas and pure gas regions. The yellow-shaded region is therefore the region of liquid + gas coexistence in the $\hat{V}$–$\hat{\tau}$ phase diagram.

On a coexistence line, following Reichl [3], we define

$$v_l = V_{\text{mix}\,l}/N_l, \qquad v_g = V_{\text{mix}\,g}/N_g, \tag{6.4a}$$

$$x_l = N_l/N, \qquad x_g = N_g/N, \qquad v_0 = V_0/N, \tag{6.4b}$$

together with the conditions

$$N_g + N_l = N \qquad V_{\text{mix}\,g} + V_{\text{mix}\,l} = V_0, \tag{6.4c}$$

where v_l and v_g are the volumes per molecule in the liquid and gas phases, respectively, x_l and x_g are the mole fractions of liquid and gas molecules, respectively, the total number of molecules is N in total volume V_0, the total volume per molecule is v_0, and N_g and N_l are the numbers of molecules in the gas and liquid phases that occupy volumes $V_{\text{mix}\,l}$ and $V_{\text{mix}\,g}$ in the two-phase region, respectively. Using equations (6.4), one can show that

$$x_l + x_g = 1, \tag{6.5a}$$

$$v_0 = x_l v_l + x_g v_g. \tag{6.5b}$$

Multiplying the left-hand side of equation (6.5*b*) by $x_l + x_g$ (= 1) and solving for x_l and x_g gives the relative number of molecules in the liquid and gas phases as

$$\frac{N_l}{N_g} = \frac{x_l}{x_g} = \frac{v_g - v_0}{v_0 - v_l}. \tag{6.6}$$

This is called the lever rule, because it is analogous to the condition for mechanical equilibrium of two children of different masses at different distances from the fulcrum of a massless seesaw (lever) under the influence of gravity.

One can solve for N_g/N and N_l/N separately as follows. Using equation (6.6) one obtains

$$N_l = \left(\frac{v_g - v_0}{v_0 - v_l}\right) N_g, \tag{6.7a}$$

$$N = N_l + N_g = \left(\frac{v_g - v_0}{v_0 - v_l} + 1\right) N_g$$

$$= \left(\frac{v_g - v_l}{v_0 - v_l}\right) N_g. \tag{6.7b}$$

Equation (6.7*b*) then gives

$$\frac{N_g}{N} = \frac{v_0 - v_l}{v_g - v_l}, \tag{6.8a}$$

$$\frac{N_l}{N} = 1 - \frac{N_g}{N} = \frac{v_g - v_0}{v_g - v_l}. \tag{6.8b}$$

A coexistence line between the pure liquid and pure gas phases in figures 6.3 and 6.6 occurs at constant pressure and temperature. Therefore the densities and volumes per molecule of coexisting gas and liquid phases are the same as at the pure gas and liquid phase end-points of the coexistence line, respectively. Thus one obtains

$$v_l \equiv \frac{V_{\text{mix}\,l}}{N_l} = \frac{V_l}{N} \qquad v_g \equiv \frac{V_{\text{mix}\,g}}{N_g} = \frac{V_g}{N}. \tag{6.9}$$

Inserting these expressions into equations (6.8) and multiplying the numerators and denominators by N/V_{c} yields

$$\frac{N_g}{N} = \frac{\hat{V}_0 - \hat{V}_l}{\hat{V}_g - \hat{V}_l}, \tag{6.10a}$$

$$\frac{N_l}{N} = \frac{\hat{V}_g - \hat{V}_0}{\hat{V}_g - \hat{V}_l}, \tag{6.10b}$$

$$\frac{N_g}{N_l} = \frac{\hat{V}_0 - \hat{V}_l}{\hat{V}_g - \hat{V}_0}, \tag{6.10c}$$

where $\hat{V}_l$ and $\hat{V}_g$ are the reduced volumes of the pure liquid and gas phases at the respective ends of a coexistence line, and $\hat{V}_0$ is the specified reduced total volume of the system. Thus equations (6.10) can be used to determine the mole fractions of gas and liquid and their ratio at a specified $\hat{V}_0$ on a coexistence line in figure 6.3 or 6.6.

For the densities of the coexisting gas and liquid phases, equations (6.9) give

$$n_{g,l} = \frac{1}{v_{g,l}} = \frac{N}{V_{g,l}}. \tag{6.11}$$

But according to equation (5.7), one has $N = n_{\mathrm{c}} V_{\mathrm{c}}$, yielding

$$n_{g,l} = \frac{n_{\mathrm{c}} V_{\mathrm{c}}}{V_{g,l}} = \frac{n_{\mathrm{c}}}{\hat{V}_{g,l}}. \tag{6.12}$$

Dividing both sides by n_{c} gives

$$\hat{n}_{g,l} \equiv \frac{n_{g,l}}{n_{\mathrm{c}}} = \frac{1}{\hat{V}_{g,l}}. \tag{6.13}$$

Thus, combined with the result that $\hat{n}_0 = 1/\hat{V}_0$ from equation (5.8), one sees that if desired, $\hat{V}_g$, $\hat{V}_l$ and/or $\hat{V}_0$ in equations (6.10) can be replaced by $1/\hat{n}_g$, $1/\hat{n}_l$ and/or $1/\hat{n}_0$, respectively.

6.4 Pressure–temperature phase diagram

The pressure–temperature phase diagram derived from the above numerical data is shown in figure 6.7. Here there are no metastable or unstable regions. The gas–liquid coexistence curve has positive slope everywhere along it and terminates in the critical point at $p = p_c$, $T = T_c$, $V = V_c$ above which the gas and liquid phases cannot be distinguished.

We obtained an analytic parameterization of the coexistence curve as follows. The $\ln(\hat{p}_X)$ versus $1/\hat{\tau}$ data were fitted by the ninth-order polynomial

$$\ln(\hat{p}_X) = \sum_{n=0}^{9} c_n (1/\hat{\tau})^n, \tag{6.14}$$

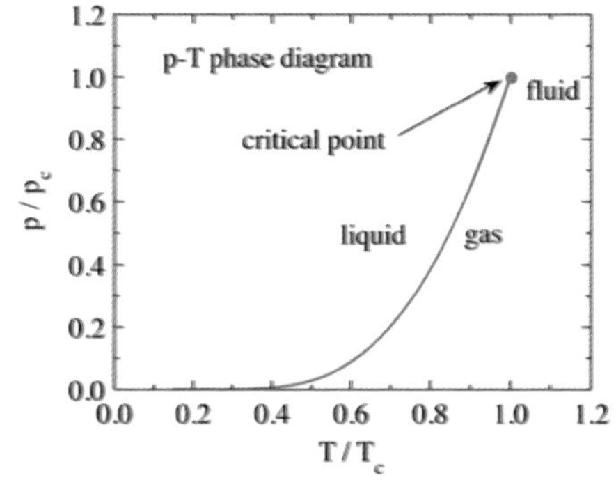

Figure 6.7. Phase diagram in the reduced pressure–temperature ($\hat{p}$–$\hat{\tau}$) plane. The coexistence curve between liquid and gas phases is indicated, which terminates at the critical point with $\hat{p} = 1, \hat{\tau} = 1, \hat{V} = 1$. On crossing the curve from left to right or top to bottom, the system transforms from pure liquid to pure gas. The liquid–gas coexistence curve is the function $\hat{p}_X(\hat{\tau})$, where $\hat{p}_X$ is defined in figure 6.2. A fit of the coexistence curve is given in equation (6.14) with coefficients c_n in table 6.1. At supercritical temperatures $\hat{\tau} > 1$, the liquid and gas phases are physically indistinguishable and the substance is termed a fluid in this region.

Table 6.1. Coefficients c_n for the parametrization of the reduced pressure $\hat{p}_X$ versus reduced temperature $\hat{\tau}$ by the ninth-order polynomial in equation (6.14) for the temperature range $0.15 \leqslant \hat{\tau} \leqslant 1$. A symbol such as 'e – 3' means '$\times 10^{-3}$'.

c_n	Value
c_0	5.664 038 35e00
c_1	−8.737 242 57e00
c_2	5.140 229 74e00
c_3	−2.925 389 42e00
c_4	1.091 088 19e00
c_5	−2.741 948 00e−1
c_6	4.599 226 54e−2
c_7	−4.928 099 27e−3
c_8	3.045 201 05e−4
c_9	−8.242 187 33e−6

where the fitted c_n coefficients are listed in table 6.1. To compare the fit with the $\hat{p}_X$ versus $\hat{\tau}$ calculated data, one exponentiates both sides of equation (6.14). The fitted $\hat{p}_X$ values agree with the calculated values to $\lesssim 0.01\%$ of $\hat{p}_X$ over the temperature range $0.15 \leqslant \hat{\tau} \leqslant 1$ of the fit.

References

[1] Reif F 1965 *Fundamentals of Statistical and Thermal Physics* (New York: McGraw-Hill)

[2] Clerk-Maxwell J 1875 On the dynamical evidence of the molecular constitution of bodies *Nature* **11** 357

[3] Reichl L E 2009 *A Modern Course in Statistical Physics* 3rd edn (Weinheim: Wiley-VCH)

David C Johnston

Chapter 7

Lekner's parametric solution of the coexistence curve and associated properties

Lekner provided an elegant and very useful alternative parametric solution for the coexistence curve in figure 6.7 and some properties associated with it that is also based on solving equations (6.1) [1]. This solution allows analytic Taylor series expansions of $\hat{p}_X$ versus $\hat{\tau}$ and associated properties to be obtained for both $\hat{\tau} \to 0$ and $\hat{\tau} \to 1$ as well as numerically exact calculations to be carried out in the intermediate temperature regime. The results obtained from Lekner's parametric solution are intrinsically more accurate than calculations using Maxwell's construction. Lekner calculated some critical exponents for $\hat{\tau} \to 1$ [1]. Berberan-Santos *et al* extended the calculations to additional properties of the vdW fluid for both $\hat{\tau} \to 0$ and $\hat{\tau} \to 1$ [2]. Here we describe and significantly extend this parametric solution and express the predictions from it in terms of our dimensionless reduced variables in equations (4.4*a*), (5.3) and (5.8).

Lekner expressed the solutions to all properties of the liquid–gas coexistence region in terms of the parameter

$$y \equiv \frac{\Delta s}{2}, \tag{7.1}$$

where Δs is the entropy difference per molecule between the gas and liquid phases in units of k_B. He also defined two functions of y as

$$f(y) = \frac{2e^y\left[(y-1)e^{2y} + y + 1\right]}{e^{4y} - 4y\,e^{2y} - 1} \tag{7.2a}$$

$$= \frac{y\cosh y - \sinh y}{\sinh y\cosh y - y}, \tag{7.2b}$$

doi:10.1088/978-1-627-05532-1ch7

$$g(y) = 1 + 2f(y)\cosh y + f^2(y). \tag{7.2c}$$

Lekner then expressed some of the properties of the coexistence region in terms of $y, f(y)$ and $g(y)$, which we augment and write in terms of the critical parameters in equation (4.4) and reduced variables in equation (5.3) with subscript X where appropriate which specifies that the quantity is associated with the coexistence region (see figures 6.1 and 6.2). Subscripts g and l refer to the coexisting gas and liquid phases, respectively. A symbol $\hat{z}$ means the value of the quantity z divided by its value z_c at the critical point. The symbols are: ΔS_X: difference in entropy between the pure gas and liquid phases at the two edges of the coexistence region in figure 6.3; $\hat{\tau}_X$: temperature on the coexistence curve in figure 6.7; $\hat{p}_X$: pressure on the coexistence curve; $\hat{V}_g$ and $\hat{V}_l$: volumes of the respective coexisting phases; and $\hat{n}_g = 1/\hat{V}_g$ and $\hat{n}_l = 1/\hat{V}_l$: number densities of the respective coexisting phases. The expressions for the gas and liquid phases are for the pure phases at the respective edges of the coexistence regions in figures 6.3 and 6.6 where all N molecules are assumed to be in the respective phase. The expressions are

$$\frac{\Delta S_X}{N k_B} \equiv \frac{S_g - S_l}{N k_B} = 2y, \tag{7.3a}$$

$$\hat{\tau}_X = \frac{27 f(y)[f(y) + \cosh y]}{4 g^2(y)}, \tag{7.3b}$$

$$\hat{p}_X = \frac{27 f^2(y)\left[1 - f^2(y)\right]}{g^2(y)}, \tag{7.3c}$$

$$\frac{d\hat{p}_X}{d\hat{\tau}} \equiv \frac{d\hat{p}_X}{d\hat{\tau}_X} = \frac{16y[y\coth(y) - 1]}{\sinh(2y) - 2y}, \tag{7.3d}$$

$$\hat{V}_g = \frac{1}{3}\left[1 + \frac{e^y}{f(y)}\right], \tag{7.3e}$$

$$\hat{V}_l = \frac{1}{3}\left[1 + \frac{e^{-y}}{f(y)}\right], \tag{7.3f}$$

$$\hat{V}_X = \hat{V}_g + \hat{V}_l = \frac{2}{3}\left[1 + \frac{\cosh y}{f(y)}\right], \tag{7.3g}$$

$$\Delta\hat{V}_X = \hat{V}_g - \hat{V}_l = \frac{2\sinh y}{3f(y)}, \tag{7.3h}$$

$$\hat{n}_g = \frac{1}{\hat{V}_g} = \frac{3f(y)}{e^y + f(y)}, \tag{7.3i}$$

$$\hat{n}_l = \frac{1}{\hat{V}_l} = \frac{3f(y)}{e^{-y} + f(y)}, \tag{7.3j}$$

$$\Delta\hat{n}_X = \hat{n}_l - \hat{n}_g = \frac{6f(y)\sinh y}{g(y)}, \tag{7.3k}$$

$$\hat{n}_{ave} = \frac{\hat{n}_l + \hat{n}_g}{2} = \frac{3f(y)[f(y) + \cosh(y)]}{g(y)}. \tag{7.3l}$$

Since $y \equiv \Delta s/2$, $\Delta s(\hat{\tau}_X = 1) = 0$ and $\Delta s(\hat{\tau}_X = 0) = \infty$ as shown below in section 7.4, the implicit variable y runs from 0 to ∞. Hence one can easily calculate the above properties including $\hat{\tau}_X$ as functions of y numerically, and then using y as an implicit parameter to evaluate the other ones as a function of $\hat{\tau}_X$ or in terms of each other. Our result for $\hat{p}_X$ versus $\hat{\tau}$ (i.e. $\hat{\tau}_X$) obtained from the parametric solution is of course the same as that already plotted using a different numerical solution in figure 6.7. However, Lekner's solution using y as the implicit parameter allows the thermodynamic properties associated with the coexistence regime to be more accurately calculated and to lower temperatures than allowed by the conventional Maxwell-construction calculations in chapter 6 using the volume as the implicit parameter.

Expressions for quantitites derived from the above fundamental ones as a function of y, along with references to the above equations originally defining them, are

$$\frac{U_g}{p_c V_c} = 4\hat{\tau}_X - 3\hat{n}_g, \quad (5.18) \tag{7.4a}$$

$$\frac{U_l}{p_c V_c} = 4\hat{\tau}_X - 3\hat{n}_l, \quad (5.18) \tag{7.4b}$$

$$\frac{U_g - U_l}{p_c V_c} = 3(\hat{n}_l - \hat{n}_g) = 3\Delta\hat{n}_X, \tag{7.4c}$$

$$\frac{H_g}{p_c V_c} = \frac{4\hat{\tau}_X(5 - \hat{n}_g)}{3 - \hat{n}_g} - 6\hat{n}_g, \quad (5.24) \tag{7.4d}$$

$$\frac{H_l}{p_c V_c} = \frac{4\hat{\tau}_X(5 - \hat{n}_l)}{3 - \hat{n}_l} - 6\hat{n}_l, \quad (5.24) \tag{7.4e}$$

$$\kappa_{\mathrm{T}g} p_{\mathrm{c}} = \frac{(3 - \hat{n}_g)^2/(6\hat{n}_g)}{4\hat{\tau}_{\mathrm{X}} - \hat{n}_g(3 - \hat{n}_g)^2}, \qquad (5.28b) \quad (7.4f)$$

$$\kappa_{\mathrm{T}l} p_{\mathrm{c}} = \frac{(3 - \hat{n}_l)^2/(6\hat{n}_l)}{4\hat{\tau}_{\mathrm{X}} - \hat{n}_l(3 - \hat{n}_l)^2}, \qquad (5.28b) \quad (7.4g)$$

$$\frac{\alpha_g \tau_{\mathrm{c}}}{k_{\mathrm{B}}} = \frac{4(3 - \hat{n}_g)/3}{4\hat{\tau}_{\mathrm{X}} - \hat{n}_g(3 - \hat{n}_g)^2}, \qquad (5.31b) \quad (7.4h)$$

$$\frac{\alpha_l \tau_{\mathrm{c}}}{k_{\mathrm{B}}} = \frac{4(3 - \hat{n}_l)/3}{4\hat{\tau}_{\mathrm{X}} - \hat{n}_l(3 - \hat{n}_l)^2}, \qquad (5.31b) \quad (7.4i)$$

$$\frac{C_{\mathrm{p}g}}{Nk_{\mathrm{B}}} = \frac{3}{2} + \frac{4\hat{\tau}_{\mathrm{X}}}{4\hat{\tau}_{\mathrm{X}} - \hat{n}_g(3 - \hat{n}_g)^2}, \qquad (5.34b) \quad (7.4j)$$

$$\frac{C_{\mathrm{p}l}}{Nk_{\mathrm{B}}} = \frac{3}{2} + \frac{4\hat{\tau}_{\mathrm{X}}}{4\hat{\tau}_{\mathrm{X}} - \hat{n}_l(3 - \hat{n}_l)^2}, \qquad (5.34b) \quad (7.4k)$$

$$\frac{L}{p_{\mathrm{c}} V_{\mathrm{c}}} = \frac{H_g - H_l}{p_{\mathrm{c}} V_{\mathrm{c}}} = \frac{16\, y\, \hat{\tau}_{\mathrm{X}}}{3}, \qquad (5.36) \quad (7.4l)$$

where U is the internal energy, H is the enthalpy, L is the latent heat (enthalpy) of vaporization on crossing the coexistence curve in figure 6.7, κ_{T} is the isothermal compressibility, α is the volume thermal expansion coefficient and C_{p} is the heat capacity at constant pressure.

Because a first-order transition occurs on crossing the coexistence curve at $\hat{p}$, $\hat{\tau} < 1$ in figure 6.7, there are discontinuities in κ_{T}, α and C_{p} on crossing the curve. One can calculate the values of these discontinuities versus $\hat{\tau}_{\mathrm{X}}$ using equations (7.4) and the parametric solutions in equations (7.3) for $\hat{n}_g$, $\hat{n}_l$ and $\hat{\tau}_{\mathrm{X}}$ with y an implicit parameter. Our analytic expressions for the discontinuities $\Delta\kappa_{\mathrm{T}} p_{\mathrm{c}} \equiv (\kappa_{\mathrm{T}g} - \kappa_{\mathrm{T}l}) p_{\mathrm{c}}$, $\Delta\alpha\tau_{\mathrm{c}}/k_{\mathrm{B}} \equiv (\alpha_g - \alpha_l)\tau_{\mathrm{c}}/k_{\mathrm{B}}$ and $\Delta C_{\mathrm{p}}/(Nk_{\mathrm{B}}) \equiv (C_{\mathrm{p}l} - C_{\mathrm{p}g})/(Nk_{\mathrm{B}})$ in terms of y derived from equations (7.3) and (7.4) are given in appendix B.

7.1 Thermodynamic behaviors on approaching the critical temperature from below

To solve for the above properties versus temperature for small deviations of $\hat{\tau}_{\mathrm{X}}$ from 1 ($y \to 0$) or 0 ($y \to \infty$) requires the solution to $y(\hat{\tau}_{\mathrm{X}})$ obtained from equation (7.3b)

in the respective limit to some order of approximation as discussed in this and the following section, respectively.

In this section the relevant quantities are the values of the parameters minus their values at the critical point. We define

$$t_0 \equiv 1 - \hat{\tau}, \qquad t_{0\mathrm{X}} \equiv 1 - \hat{\tau}_\mathrm{X}, \tag{7.5}$$

which are positive for $\hat{\tau}$, $\hat{\tau}_\mathrm{X} < 1$. Taylor expanding equation (7.3*b*) to 6th order in y gives

$$t_{0\mathrm{X}} = \frac{y^2}{9} - \frac{y^4}{75} + \frac{946y^6}{637\,875}. \tag{7.6a}$$

Solving for $y(t_{0\mathrm{X}})$ to lowest orders gives

$$y = 3\,t_{0\mathrm{X}}^{1/2} + \frac{81\,t_{0\mathrm{X}}^{3/2}}{50} + \frac{50\,403\,t_{0\mathrm{X}}^{5/2}}{35\,000}. \tag{7.6b}$$

Taylor expanding equations (7.3) about $y = 0$, substituting equation (7.6*b*) into these Taylor series expansions and simplifying gives the $y \to 0$ and $t_{0\mathrm{X}} \to 0$ behaviors of the quantities in equations (7.3) to lowest orders as

$$\frac{\Delta S_\mathrm{X}}{Nk_\mathrm{B}} \equiv 2y = 6\,t_{0\mathrm{X}}^{1/2} + \frac{81\,t_{0\mathrm{X}}^{3/2}}{25} + \frac{50\,403\,t_{0\mathrm{X}}^{5/2}}{17\,500}, \tag{7.7a}$$

$$\frac{L}{p_\mathrm{c}V_\mathrm{c}} = \frac{16\hat{\tau}_\mathrm{X}(y)y}{3} = \frac{16y}{3} - \frac{16y^3}{27} + \frac{16y^5}{225} = 16\,t_{0\mathrm{X}}^{1/2} - \frac{184\,t_{0\mathrm{X}}^{3/2}}{25} - \frac{4198\,t_{0\mathrm{X}}^{5/2}}{4375}, \tag{7.7b}$$

$$p_{0\mathrm{X}} \equiv \hat{p}_\mathrm{X} - 1 = -\frac{4y^2}{9} + \frac{76y^4}{675} - \frac{13\,672y^6}{637\,875} = -4\,t_{0\mathrm{X}} + \frac{24\,t_{0\mathrm{X}}^2}{5} - \frac{816\,t_{0\mathrm{X}}^3}{875}, \tag{7.7c}$$

$$\frac{\mathrm{d}\hat{p}_\mathrm{X}}{\mathrm{d}\hat{\tau}_\mathrm{X}} = 4 - \frac{16y^2}{15} + \frac{256y^4}{1575} - \frac{64y^6}{3375} = 4 - \frac{48\,t_{0\mathrm{X}}}{5} + \frac{2448\,t_{0\mathrm{X}}^2}{875} + \frac{56\,832\,t_{0\mathrm{X}}^3}{21\,875}, \tag{7.7d}$$

$$v_{0g} \equiv \hat{V}_g - 1 = \frac{2y}{3} + \frac{2y^2}{5} + \frac{8y^3}{45} = 2\,t_{0\mathrm{X}}^{1/2} + \frac{18\,t_{0\mathrm{X}}}{5} + \frac{147\,t_{0\mathrm{X}}^{3/2}}{25}, \tag{7.7e}$$

$$v_{0l} \equiv \hat{V}_l - 1 = -\frac{2y}{3} + \frac{2y^2}{5} - \frac{8y^3}{45} = -2\,t_{0\mathrm{X}}^{1/2} + \frac{18\,t_{0\mathrm{X}}}{5} - \frac{147\,t_{0\mathrm{X}}^{3/2}}{25}, \tag{7.7f}$$

$$v_{0\mathrm{X}} \equiv v_{0g} - v_{0l} = \frac{4y}{3} + \frac{16y^3}{45} + \frac{32y^5}{1575} = 4\,t_{0\mathrm{X}}^{1/2} + \frac{294\,t_{0\mathrm{X}}^{3/2}}{25} + \frac{196\,081\,t_{0\mathrm{X}}^{5/2}}{8750}, \tag{7.7g}$$

$$\hat{V}_{\rm X} = 2 + v_{0g} + v_{0l} = 2 + \frac{4y^2}{5} + \frac{68y^4}{525} + \frac{32y^5}{1575}$$
$$= 2 + \frac{36\, t_{0\rm X}}{5} + \frac{15\,984\, {t_{0\rm X}}^2}{875} + \frac{864\, {t_{0\rm X}}^{5/2}}{175}, \tag{7.7h}$$

$$n_{0g} \equiv \hat{n}_g - 1 = -\frac{2y}{3} + \frac{2y^2}{45} + \frac{8y^3}{135} = -2\, {t_{0\rm X}}^{1/2} + \frac{2\, t_{0\rm X}}{5} + \frac{13\, {t_{0\rm X}}^{3/2}}{25}, \tag{7.7i}$$

$$n_{0l} \equiv \hat{n}_l - 1 = \frac{2y}{3} + \frac{2y^2}{45} - \frac{8y^3}{135} = 2\, {t_{0\rm X}}^{1/2} + \frac{2\, t_{0\rm X}}{5} - \frac{13\, {t_{0\rm X}}^{3/2}}{25}, \tag{7.7j}$$

$$\Delta\hat{n}_{\rm X} \equiv \hat{n}_l - \hat{n}_g = \frac{4y}{3} - \frac{16y^3}{135} + \frac{544y^5}{42\,525} = 4\, {t_{0\rm X}}^{1/2} - \frac{26\, {t_{0\rm X}}^{3/2}}{25} - \frac{1359\, {t_{0\rm X}}^{5/2}}{8750}, \tag{7.7k}$$

$$\hat{n}_{\rm ave\,X} \equiv \frac{\hat{n}_l + \hat{n}_g}{2} = 1 + \frac{2y^2}{45} - \frac{2y^4}{567} + \frac{4y^6}{18\,225} = 1 + \frac{2\, t_{0\rm X}}{5} + \frac{128\, {t_{0\rm X}}^2}{875} + \frac{136\, {t_{0\rm X}}^3}{3125}. \tag{7.7l}$$

In these expressions, it is important to remember the definition $t_{0\rm X} \equiv 1 - \hat{\tau}_{\rm X}$ in equations (7.5). Thus, $t_{0\rm X}$ increases as $\hat{\tau}_{\rm X}$ decreases below the critical temperature. The leading expression in the last equality of each equation is the asymptotic critical behavior of the quantity as $\hat{\tau} \to 1^-$, as further discussed in chapter 8 below.

7.2 Thermodynamic behaviors for temperatures approaching zero

Expanding the hyperbolic functions in the expression for $\hat{\tau}_{\rm X}(y)$ in equation (7.3*b*) into their constituent exponentials gives

$$\hat{\tau}_{\rm X} = \frac{27\left[1 + (y-1)e^{2y} + y\right]\left(e^{4y} - 4y\, e^{2y} - 1\right)^2}{4\left(e^{2y} - 1\right)\left[(2y-1)e^{4y} + \left(2 - 4y^2\right)e^{2y} - 2y - 1\right]^2}. \tag{7.8}$$

The method of determining the behavior of $\hat{\tau}_{\rm X}(y)$ at low temperatures where $y \to \infty$ is the same for all thermodynamic variables and functions. The behaviors of the numerator and denominator on the right-hand side of equation (7.8) are dominated by the respective exponential with the highest power of y. Retaining only those exponentials and their prefactors, equation (7.8) becomes

$$\hat{\tau}_{\rm X} = \frac{27\left[(y-1)e^{2y}\right]e^{8y}}{4e^{2y}\left[(2y-1)e^{4y}\right]^2} = \frac{27(y-1)}{4(2y-1)^2}. \tag{7.9}$$

In this case, the exponentials cancel out but for other quantities they do not. Taylor expanding the expression on the far right of equation (7.9) in powers of $1/y$ to order $1/y^4$ gives

$$\hat{\tau}_X(y) = \frac{27}{16y} - \frac{27}{64y^3} - \frac{27}{64y^4} \qquad (y \to \infty). \tag{7.10}$$

Interestingly, the y^{-2} term is zero. Solving for $y(\hat{\tau}_X)$ to order $\hat{\tau}_X^2$ gives

$$y = \frac{27}{16\,\hat{\tau}_X} - \frac{4\,\hat{\tau}_X}{27} - \frac{64\,\hat{\tau}_X^2}{729} \qquad (\hat{\tau}_X \to 0), \tag{7.11}$$

where here the $\hat{\tau}_X^0$ term is zero. The entropy difference and latent heat for $\hat{\tau}_X \to 0$ are obtained by substituting equation (7.11) into equations (7.3*a*) and (7.4*l*), respectively. The low-temperature limiting behaviors of the other functions versus y are obtained as above for $\hat{\tau}_X(y)$. If there is an exponential still present after the above reduction, it is of course retained. In that case, only the leading order term of $y(\hat{\tau}_X)$ is inserted into the argument of the exponential, equation (7.11) is inserted for y in the exponential prefactor and then a power series in $1/\hat{\tau}_X$ is obtained for the prefactor. The results for the low-order terms for $1/y \to 0$ and $\hat{\tau}_X \to 0$ for the quantities in equations (7.3) are

$$\frac{\Delta S_X}{Nk_B} = 2\,y = \frac{27}{8\,\hat{\tau}_X} - \frac{8\,\hat{\tau}_X}{27} - \frac{128\,\hat{\tau}_X^2}{729}, \tag{7.12a}$$

$$\frac{L}{p_c V_c} = \frac{16\,\hat{\tau}_X\,y}{3} = 9 - \frac{64\,\hat{\tau}_X^2}{81} - \frac{1024\,\hat{\tau}_X^3}{2187} - \frac{8192\,\hat{\tau}_X^4}{59\,049}, \tag{7.12b}$$

$$\hat{p}_X = \frac{108(y-1)^2 e^{-2y}}{(2y-1)^2} = \left(\frac{1\,594\,323}{256\,\hat{\tau}_X^3} + \frac{98\,415}{64\,\hat{\tau}_X^2} - \frac{5103}{4\,\hat{\tau}_X}\right) e^{\frac{-27}{8\,\hat{\tau}_X}}, \tag{7.12c}$$

$$\hat{V}_g = \frac{e^{2y}}{6(y-1)} = \left(\frac{8\,\hat{\tau}_X}{81} + \frac{128\,\hat{\tau}_X^2}{2187} + \frac{2560\,\hat{\tau}_X^3}{59\,049}\right) e^{\frac{27}{8\,\hat{\tau}_X}}, \tag{7.12d}$$

$$\hat{V}_l = \frac{2y-1}{6(y-1)} = \frac{1}{3} + \frac{8\,\hat{\tau}_X}{81} + \frac{128\,\hat{\tau}_X^2}{2187} + \frac{2560\,\hat{\tau}_X^3}{59\,049}, \tag{7.12e}$$

$$\hat{n}_g = \frac{1}{\hat{V}_g} = 6(y-1)e^{-2y} = 6\left(\frac{27}{16\,\hat{\tau}_X} - 1 - \frac{4\,\hat{\tau}_X}{27} - \frac{64\,\hat{\tau}_X^2}{729}\right) e^{\frac{-27}{8\,\hat{\tau}_X}}, \tag{7.12f}$$

$$\hat{n}_l = \frac{1}{\hat{V}_l} = \frac{6(y-1)}{2y-1} = 3 - \frac{8\,\hat{\tau}_X}{9} + \frac{64\,\hat{\tau}_X^2}{243} - \frac{1024\,\hat{\tau}_X^3}{6561}, \tag{7.12g}$$

$$\hat{n}_{\text{ave X}} = \frac{\hat{n}_l}{2} = \frac{3(y-1)}{2y-1} = \frac{3}{2} - \frac{4\,\hat{\tau}_X}{9} + \frac{32\,\hat{\tau}_X^2}{243} - \frac{512\,\hat{\tau}_X^3}{6561}. \tag{7.12h}$$

In equation (7.12*h*), we used the fact that $\hat{n}_g(\hat{\tau}_X)$ approaches zero exponentially for $\hat{\tau}_X \to 0$ instead of as a power law as does $\hat{n}_l(\hat{\tau}_X)$.

7.3 Coexisting liquid and gas densities, transition order parameter and temperature–density phase diagram

The densities $\hat{n}_g$ and $\hat{n}_l$ of coexisting gas and liquid phases obtained from equations (7.3*i*) and (7.3*j*), respectively, together with equation (7.3*b*) are plotted versus reduced temperature $\hat{\tau} = T/T_c$ in figure 7.1(*a*). At the critical temperature they become the same. The difference $\Delta\hat{n}_X \equiv \hat{n}_l - \hat{n}_g$ is the order parameter of the gas–liquid transition and is plotted versus $\hat{\tau}$ in figure 7.1(*b*).

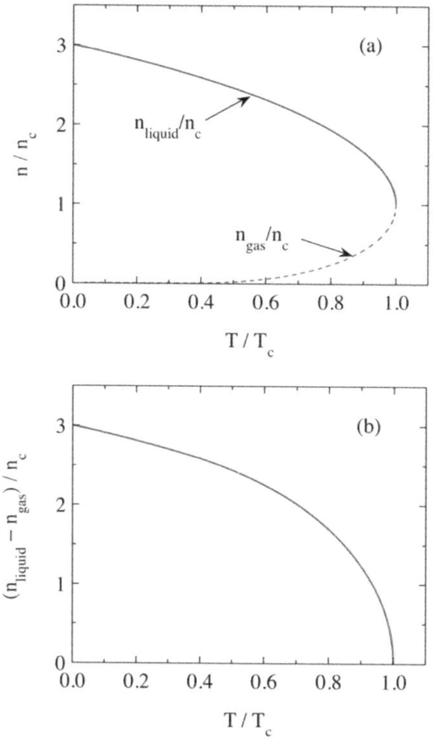

Figure 7.1. (*a*) Reduced densities $\hat{n}_g = n_g/n_c$ and $\hat{n}_l = n_l/n_c$ of coexisting gas and liquid phases versus reduced temperature $\hat{\tau} = T/T_c$ from equations (7.3*i*) and (7.3*j*), respectively, together with (7.3*b*). (*b*) Order parameter $\Delta\hat{n}_X \equiv \hat{n}_l - \hat{n}_g$ of the liquid–gas phase transition versus $\hat{\tau}$ obtained from the data in (*a*).

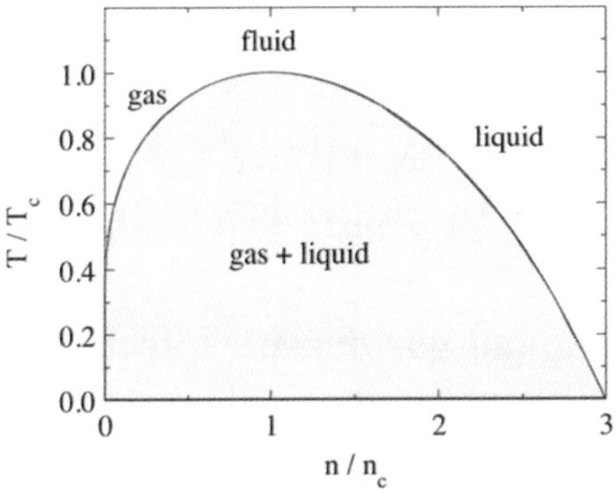

Figure 7.2. Reduced temperature $\hat{\tau} = T/T_c$ versus reduced density $\hat{n} = n/n_c$ phase diagram for the vdW fluid, constructed by reversing the axes of figure 7.1(*a*). The maximum reduced density the system can have is $\hat{n} = 3$, at which there is no free volume left in which the molecules can move.

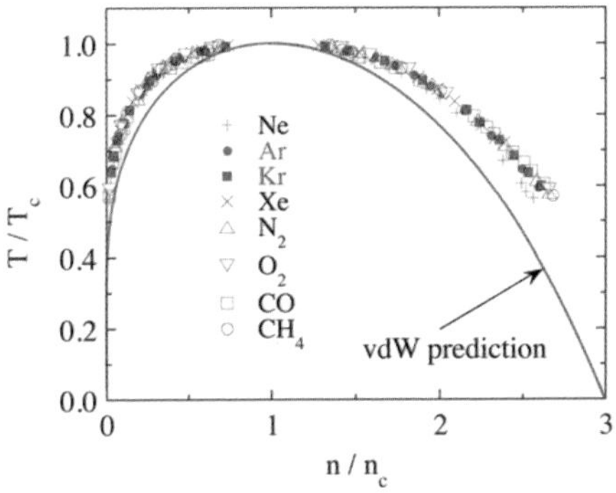

Figure 7.3. Reduced temperature $\hat{\tau} = T/T_c$ versus reduced density $\hat{n} = n/n_c$ of the coexisting gas and liquid phases of eight different fluids [3]. Also shown as the solid curve is the prediction for the vdW fluid from figure 7.2. The experimental data follow a law of corresponding states [3], but the one predicted for the vdW fluid does not accurately describe the data.

Data such as in figure 7.1(*a*) are often plotted with reversed axes, yielding the temperature–density phase diagram [3] in figure 7.2. The phase diagram and associated temperature dependences of the coexisting densities of the liquid and gas phases experimentally determined for eight different gases are shown in figure 7.3, along with the prediction for the vdW fluid from figure 7.2. The experimental data were digitized from Fig. 2 of [3]. Interestingly, the experimental data follow a law of corresponding states [3], although that law does not quantitatively agree with the one predicted for the vdW fluid.

A comparison of the high- and low-temperature limits of the average density $\hat{n}_{\rm ave}$ in equations (7.7*l*) and (7.12*h*), respectively, of the coexisting gas and liquid phases shows that $\hat{n}_{\rm ave}$ is not a rectilinear function of temperature, which was noted by Lekner [1]. Shown in figure 7.4 is a plot of $\hat{n}_{\rm ave}$ versus $\hat{\tau}$ obtained from equations (7.3*b*) and (7.3*l*), which shows an S-shaped behavior.

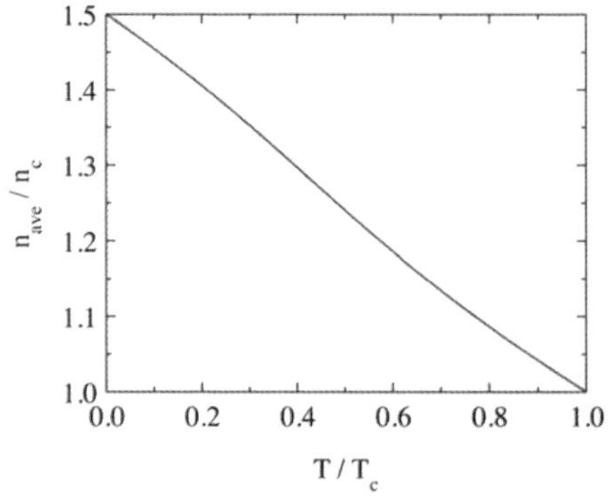

Figure 7.4. Reduced average number density of coexisting liquid and gas phases $\hat{n}_{\text{ave}} = (\hat{n}_l + \hat{n}_g)/2$ versus reduced temperature $\hat{\tau} = T/T_c$ obtained from equations (7.3*b*) and (7.3*l*). The curve has an S-shape and hence is not rectilinear.

7.4 Latent heat and entropy of vaporization

The normalized latent heat (or enthalpy) of vaporization $L/(p_c V_c)$ on crossing the liquid–gas coexistence curve in figure 6.7 is obtained parametrically versus $\hat{\tau}_X$ from equations (7.3*b*) and (7.4*l*) and is plotted in figure 7.5(*c*). The low-temperature behavior agrees with the prediction in equation (7.12*b*). From figure 7.5(*c*), one sees that $L \to 0$ as $T \to T_c^-$, which is required because at temperatures at and above the critical temperature, the liquid and gas phases are no longer physically distinguishable. The normalized entropy of vaporization $\Delta S_X/(Nk_B)$ is obtained from equations (7.3*a*) and (7.3*b*) and is plotted versus $\hat{\tau}_X$ in figure 7.6. The entropy difference is seen to diverge for $\hat{\tau}_X \to 0$, in agreement with equation (7.12*a*).

From the reduced $\hat{p}_X(\hat{\tau})$ data and information about the change in reduced volume $\Delta\hat{V}_X = \hat{V}_{\text{gas}} - \hat{V}_{\text{liquid}}$ across the coexistence line obtained from numerical calculations, one can also determine L using the Clausius–Clapeyron equation

$$\frac{\mathrm{d}p_X}{\mathrm{d}T} = \frac{L}{T\Delta V_X}, \tag{7.13a}$$

or

$$L = T\Delta V_X \frac{\mathrm{d}p_X}{\mathrm{d}T}. \tag{7.13b}$$

One can write equation (7.3*b*) in terms of the reduced variables in equations (5.3) as

$$\frac{L}{p_c V_c} = \hat{\tau}\Delta\hat{V}_X \frac{\mathrm{d}\hat{p}_X}{\mathrm{d}\hat{\tau}}. \tag{7.13c}$$

The variations of $\mathrm{d}\hat{p}_X/\mathrm{d}\hat{\tau}$ versus $\hat{\tau}$ obtained from equations (7.3*b*) and (7.3*d*) and $\Delta\hat{V}_X$ from equations (7.3*b*) and (7.3*h*) versus $\hat{\tau}$ are shown in figures 7.5(*a*) and 7.5(*b*), respectively. These behaviors when inserted into equation (7.13*c*) give the same

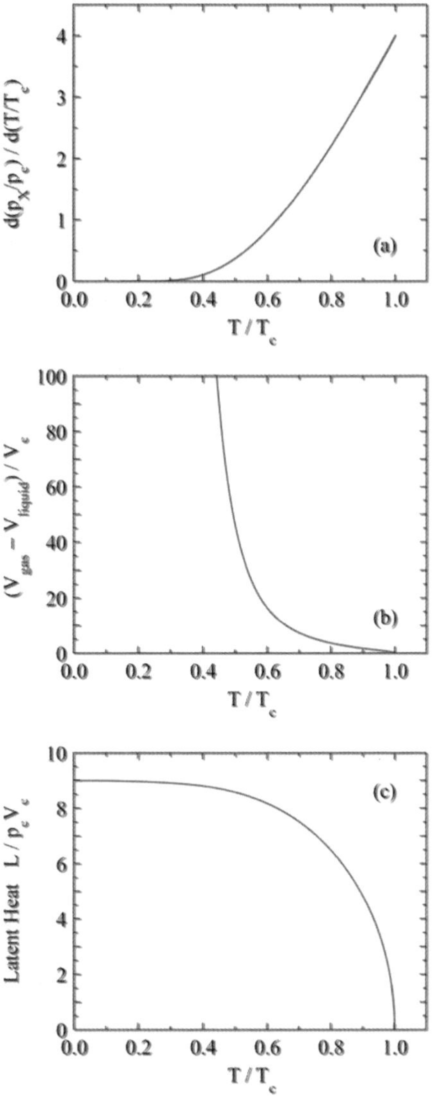

Figure 7.5. (*a*) The derivative $d\hat{p}_X/d\hat{\tau}$ versus $\hat{\tau}$ of the liquid–gas coexistence curve in figure 6.7. (*b*) Difference in normalized volume $\Delta\hat{V}_X = \Delta V_X/V_c \equiv (V_{gas} - V_{liquid})/V_c$ on crossing the p–T gas–liquid coexistence line in figure 6.7 versus $\hat{\tau}$. $\Delta\hat{V}_X$ diverges for $\hat{\tau} \to 0$, consistent with equations (7.12*d*) and (7.12*e*). (*c*) Latent heat L versus $\hat{\tau}$ obtained from either equation (7.13*c*) using data as in panels (*a*) and (*b*) or from the parametric solution in equations (7.3*b*) and (7.4*l*). Both calculations give identical results to within their respective numerical accuracies.

$L/(p_c V_c)$ versus $\hat{\tau}_X$ behavior as already obtained from Lekner's parametric solution in figure 7.5(*c*).

The entropy change $\Delta S_X \equiv S_{gas} - S_{liquid}$ on moving left to right across the p–T liquid–gas coexistence curve in figure 6.7 is given in reduced units by equation (5.36) as

$$\frac{\Delta S_X(\hat{\tau})}{Nk_B} = \frac{3}{8\,\hat{\tau}}\left[\frac{L(\hat{\tau})}{p_c V_c}\right]. \tag{7.14}$$

The quantity in square brackets on the right-hand side is already plotted in figure 7.5(*c*). Using these data and (7.14) yields $\Delta S_X/(Nk_B)$ versus $\hat{\tau}$ which is the same as that already plotted using Lekner's solution in figure 7.6. The entropy change goes to zero at the critical point because gas and liquid phases cannot be distinguished at and above the critical temperature. From figure 7.5(*a*), the derivative $\mathrm{d}\hat{p}_X/\mathrm{d}\hat{\tau}$ shows no critical divergence. Therefore, according to equation (7.13*c*), ΔS_X shows the same critical behavior for $T \to T_c$ as does $\Delta \hat{V}_X$ (or $\Delta \hat{n}_X$, see chapter 8 below).

Since the latent heat becomes constant at low temperatures according to figure 7.5(*c*), ΔS_X diverges to ∞ as $T \to 0$ according to equation (7.14), as seen in figure 7.6. This divergence violates the third law of thermodynamics which states that the entropy of a system must tend to a constant value (usually zero) as $T \to 0$. This behavior demonstrates that like the ideal gas, the vdW fluid is classical. This means that the predictions of the thermodynamic properties for either system are only valid in the large-volume classical regime where the number density N/V of the system is much less than the quantum concentration n_Q. Furthermore, the triple points of materials, where solid, gas and liquid coexist, typically occur at $T/T_c \sim 1/2$, so this also limits the temperature range over which the vdW theory is applicable to real fluids. However, the study of the vdW fluid at lower temperatures is still of theoretical interest.

Representative values for the above properties associated with the coexistence curve that we calculated using the parametric equations (7.3) and (7.4) are listed in table A.2 in appendix A.

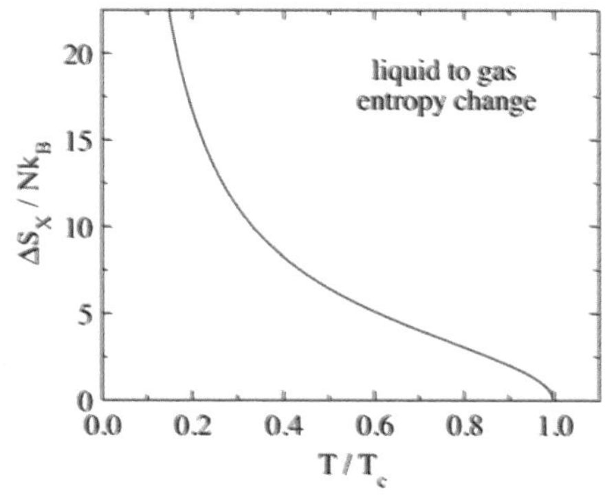

Figure 7.6. Reduced entropy change at the first-order liquid–gas transition $\Delta S_X/(Nk_B) = (S_{gas} - S_{liquid})/(Nk_B)$ versus reduced temperature $\hat{\tau} = T/T_c$ on crossing the coexistence curve in figure 6.7. The data plotted can be obtained from either the data in figure 7.5(*c*) using equation (7.14) or from the parametric solution in equations (7.3*a*) and (7.3*b*). According to equation (7.14), ΔS_X diverges to ∞ as $T \to 0$ because the latent heat of vaporization becomes constant at low temperatures according to figure 7.5(*c*) and equation (7.12*b*).

7.5 Heat capacity at constant volume versus temperature within the liquid–gas coexistence region along an isochoric path

We consider $C_V(T < T_c)$ within the coexistence region separately from the above thermodynamic properties associated with gas–liquid coexistence because it is more involved. Apparently, only the limiting behavior $C_V(T \to T_c^-)$ has been calculated versus temperature for a path along the critical isotherm; this calculation was carried out in order to obtain the critical exponent α' and amplitude a' as discussed below in chapter 8. We find that C_V depends significantly on T for $T < T_c$. It may seem strange that C_V is temperature-dependent in the coexistence region, which consists of a mixture of gas and liquid phases, each of which as a pure phase has the same T-independent $C_V/(Nk_B) = 3/2$ per molecule. However, a latent heat is associated with converting pure liquid to pure gas and therein lies the reason for the T dependence of C_V within the coexistence region. In particular, figure 7.1(a) shows that the ratio of gas to liquid molecules depends on T for $T < T_c$ and figure 7.6 shows that the entropy difference per molecule between the coexisting liquid and gas phases also depends on temperature for $T < T_c$. The combination of these two behaviors results in a C_V that depends on T in the coexistence region. Here we present a straightforward derivation of an analytic $C_V(T)$ along a chosen isochoric path with fixed reduced volume $\hat{V}_0$ that is exact within the entire liquid–gas coexistence region of figure 6.3. Our solution is based on Lekner's analytic parameterization of the coexistence properties as discussed above.

In the coexistence region containing both gas (g) and liquid (l) phases, the separate and total internal energies are given by the basic definition of the internal energy of a vdW fluid in equation (4.6) as

$$\begin{aligned} U_g &= \frac{3}{2}N_g\tau - a\frac{N_g^2}{V_{\text{mix}\,g}}, \\ U_l &= \frac{3}{2}N_l\tau - a\frac{N_l^2}{V_{\text{mix}\,l}}, \\ U = U_g + U_l &= \frac{3}{2}N\tau - a\left(\frac{N_g^2}{V_{\text{mix}\,g}} + \frac{N_l^2}{V_{\text{mix}\,l}}\right) \end{aligned} \tag{7.15}$$

where U is the total internal energy of the liquid and gas mixture, U_g and U_l are the internal energies of the gas and liquid phases, respectively, N_g and N_l are the numbers of molecules in the gas and liquid phases, respectively, and $V_{\text{mix}\,g}$ and $V_{\text{mix}\,l}$ are the volumes occupied by the mixed ('mix') gas and liquid phases, respectively. The constraints for calculating C_V versus T along an isochore are that N and the isochore volume V_0 are fixed with

$$\begin{aligned} N &= N_g + N_l, & \mathrm{d}N_g &= -\mathrm{d}N_l, \\ V_0 &= V_{\text{mix}\,g} + V_{\text{mix}\,l}, & \mathrm{d}V_{\text{mix}\,g} &= -\mathrm{d}V_{\text{mix}\,l}. \end{aligned} \tag{7.16}$$

Using these constraints, equation (7.15), and equation (4.7) adapted to the gas and liquid phases, the heat capacity at constant volume along an isochore is

$$\frac{C_{\rm V}}{Nk_{\rm B}} = \left(\frac{\partial U}{\partial T}\right)_{N,V_0}$$
$$= \frac{3}{2} - a\left[2(n_l - n_g)\frac{\mathrm{d}(N_l/N)}{\mathrm{d}\tau} - \frac{n_l^2 - n_g^2}{N}\frac{\mathrm{d}V_{\mathrm{mix}\,l}}{\mathrm{d}\tau}\right], \tag{7.17}$$

where we substituted the definitions $\tau = k_{\rm B}T$ and $n_{g,l} = N_{g,l}/V_{\mathrm{mix}\,g,l}$.

To calculate $\mathrm{d}V_{\mathrm{mix}\,l}/\mathrm{d}\tau$ in equation (7.17) we use the first of equations (6.9) to write

$$V_{\mathrm{mix}\,l} = \left(\frac{N_l}{N}\right)V_l, \tag{7.18}$$

where V_l is the volume of the liquid at the edge of a coexistence line in figures 6.3 or 6.6 as defined above in Lekner's solution of the coexistence curve. Both N_l/N and V_l change with increasing T, so one obtains

$$\frac{\mathrm{d}V_{\mathrm{mix}\,l}}{\mathrm{d}\tau} = V_l\frac{\mathrm{d}(N_l/N)}{\mathrm{d}\tau} + \left(\frac{N_l}{N}\right)\frac{\mathrm{d}V_l}{\mathrm{d}\tau}. \tag{7.19}$$

Inserting equation (7.19) into equation (7.17) and using the relations $N = n_{\rm c}V_{\rm c}$ from equation (5.7) and $\hat{V}_l \equiv V_l/V_{\rm c}$ gives

$$\frac{C_{\rm V}}{Nk_{\rm B}} = \frac{3}{2} - a\left\{2(n_l - n_g)\frac{\mathrm{d}(N_l/N)}{\mathrm{d}\tau} - \frac{n_l^2 - n_g^2}{n_{\rm c}}\left[\hat{V}_l\frac{\mathrm{d}(N_l/N)}{\mathrm{d}\tau} + \left(\frac{N_l}{N}\right)\frac{\mathrm{d}\hat{V}_l}{\mathrm{d}\tau}\right]\right\}. \tag{7.20}$$

Now we convert to reduced density and temperature units as already done for the volume $\hat{V}$, yielding

$$\frac{C_{\rm V}}{Nk_{\rm B}} = \frac{3}{2} - \frac{9}{8}\left\{2(\hat{n}_l - \hat{n}_g)\frac{\mathrm{d}(N_l/N)}{\mathrm{d}\hat{\tau}} - (\hat{n}_l^2 - \hat{n}_g^2)\left[\hat{V}_l\frac{\mathrm{d}(N_l/N)}{\mathrm{d}\hat{\tau}} + \left(\frac{N_l}{N}\right)\frac{\mathrm{d}\hat{V}_l}{\mathrm{d}\hat{\tau}}\right]\right\} \tag{7.21a}$$

where we used the equality $an_{\rm c}/\tau_{\rm c} = 9/8$ obtained from the expression for a in equations (4.4*d*) and for $n_{\rm c}$ in equation (5.6*b*). For clarity, the expression for N_l/N from equation (6.10*b*) is

$$\frac{N_l}{N} = \frac{\hat{V}_g - \hat{V}_0}{\hat{V}_g - \hat{V}_l}. \tag{7.21b}$$

We use Lekner's parametrization where the implicit parameter is y, the quantities $\hat{\tau} \equiv \hat{\tau}_{\rm X}$, $\hat{n}_l$, $\hat{n}_g$, $\hat{V}_l$ and $\hat{V}_g$ in equations (7.21) are all functions of y as given above in equations (7.3). Thus the derivatives with respect to temperature in equation (7.21*a*) within this parametrization are calculated using

$$\frac{\mathrm{d}}{\mathrm{d}\hat{\tau}} = \left[\frac{1}{\mathrm{d}\hat{\tau}(y)/\mathrm{d}y}\right]\frac{\mathrm{d}}{\mathrm{d}y}. \tag{7.21c}$$

Because N_l/N in equation (7.21*b*) is a function of the chosen isochore value $\hat{V}_0$ along which C_V is calculated, the functional dependence of C_V in equation (7.21*a*) is

$$\frac{C_V\left(y,\ \hat{V}_0\right)}{Nk_B}.$$

To calculate and plot $C_V/(Nk_B)$ versus $\hat{\tau}$ one makes a column of $\hat{\tau}$ values for a list of y values using equation (7.3*b*) and another column of C_V versus the same list of y values using equations (7.21), from which one obtains C_V versus $\hat{\tau}$ with y as an implicit parameter. Once the two-phase region is exited into the pure gas, liquid or undifferentiated fluid phase regions with increasing temperature, the $C_V/(Nk_B)$ is given by the result in equation (4.7) that $C_V/(Nk_B) = 3/2$ for these cases. For $\hat{V}_0 = 1$, this transition of course occurs for $\hat{\tau} = 1$. However, for $\hat{V}_0 \neq 1$, the change of the calculation from equations (7.21) to (4.7) as the transition temperature in figures 6.3 and 6.6 is traversed for the chosen value of $\hat{V}_0$ is lower. For $\hat{V}_0 < 1$, the transition temperature is the temperature at which $\hat{V}_0 = \hat{V}_G$ in figure 6.2 and table A.1 of the appendix, whereas for $\hat{V}_0 > 1$ the transition temperature is the temperature at which $\hat{V}_0 = \hat{V}_C$ in figure 6.2 and table A.1.

The $C_V(\hat{\tau})/(Nk_B)$ for the vdW fluid was calculated using equations (7.3*b*) and (7.21) for representative isochoric paths with $\hat{V}_0 = 0.5$, 1 and 2. The results are shown in figure 7.7. At low temperatures, $C_V/(Nk_B)$ approaches the nonzero value

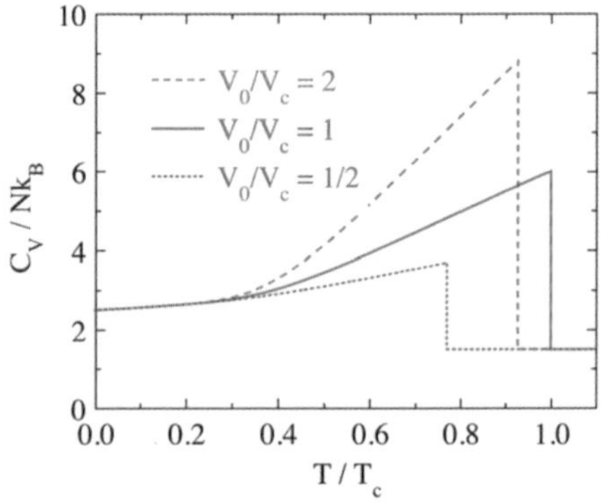

Figure 7.7. Normalized heat capacity at constant volume $C_V/(Nk_B)$ versus reduced temperature $\hat{\tau} = T/T_c$ with paths along isochores with reduced volumes $\hat{V}_0 \equiv V_0/V_c = 1/2$, 1 and 2. The mixed gas–liquid phase to pure phase transition temperatures for $\hat{V}_0 = 1/2$ and 2 are lower than for $\hat{V}_0 = 1$ (see figures 6.3 and 6.6). The transitions are second-order transitions with no latent heat as explained in the text.

of 5/2 for each value of $\hat{V}_0$, corresponding to a divergence in the entropy of the system and a violation of the third law of thermodynamics, typical of classical theories of thermodynamic properties. The $C_V(\hat{\tau})/(Nk_B)$ data for each value of $\hat{V}_0$ show finite but different values at the respective transition temperature from the mixed-phase region to a single-phase region. Representative values of $C_V/(Nk_B)$ versus y and $\hat{\tau}$ for $\hat{V}_0 = 0.5$, 1 and 2 are given in table A.3 in appendix A.

The shape of the $C_V(T)$ curves in figure 7.7 is typical of systems showing second-order mean-field phase transitions, such as the magnetic heat capacity at zero applied magnetic field C_{mag} versus T of local moment Heisenberg ferromagnets and antiferromagnets in the Weiss molecular field approximation [4]. However, $C_{mag}(T \to 0) = 0$ is obtained for spin systems with finite spin quantum numbers S, and only for the classical limit of infinite spin does one obtain $C_{mag}(T \to 0) \neq 0$.

For the vdW fluid, the transition is second order for $\hat{V}_0 = 1$ because the latent heat of vaporization is zero at T_c as seen in figure 7.5(*c*). However, for $\hat{V}_0 = 0.5$ and 2 where the transition temperature is lower than the critical temperature T_c, one must remember that the latent heat versus T in figure 7.5(*c*) is the heat required to completely convert pure liquid to pure gas at a given temperature. On the other hand, referring to the p–V isotherms in figure 6.3, one sees that as one traverses a vertical isochoric path with, say, $\hat{V}_0 = 2$, from low temperatures to the temperature of the boundary of the coexistence region, according to the lever rule in section 6.3 the coexisting liquid is continuously converted into gas until at the boundary of the coexistence region there is no liquid left to convert. Similarly, again referring to figure 6.3, for $\hat{V}_0 < 1$ the amount of gas that gets converted into liquid goes continuously to zero as the transition temperature is approached at fixed volume. Therefore, irrespective of the value of $\hat{V}_0$ for the isochoric path in the coexistence region, the transition from the mixed phase to a pure phase with increasing temperature is second order with no discontinuity in the relative amounts of gas and liquid on traversing the transition, and hence no latent heat.

The analytic expression for $C_V(y)$ for the important path along the critical isochore $\hat{V} = 1$ obtained from equations (7.21) is given in appendix C. The limiting behaviors of $C_V/(Nk_B)$ for $\hat{\tau} \to 1^-$ $(y \to 0)$ and $\hat{\tau} \to 0$ $(y \to \infty)$ have been determined for this path from this expression. Taylor series expansions of $C_V(y)/(Nk_B)$ for $y \to 0$ and $y \to \infty$ were obtained analytically from equations (7.21) as described for other functions in sections 7.1 and 7.2 above. Then the $\hat{\tau}$ dependences were obtained by converting the y parameter to $\hat{\tau}$ using equations (7.6*b*) and (7.10), respectively. The results to lowest orders are

$$\begin{aligned}\frac{C_V(\hat{\tau} \to 1^-)}{Nk_B} &= 6 - \frac{14}{25}y^2 + \frac{796}{13\,125}y^4 - \frac{41\,024}{6\,890\,625}y^6 \\ &= 6 - \frac{126}{25}t_0 - \frac{2322}{4375}t_0^2 - \frac{1152}{30\,625}t_0^3,\end{aligned} \qquad (7.22a)$$

$$\frac{C_V(\hat{\tau} \to 0)}{Nk_B} = \frac{10y - 11}{4y - 6}$$
$$= \frac{5}{2} + \frac{16}{27}\hat{\tau} + \frac{128}{243}\hat{\tau}^2 + \frac{10\,240}{19\,683}\hat{\tau}^3, \quad (7.22b)$$

where $t_0 \equiv 1 - \hat{\tau}$ according to equation (7.5). Thus, the heat capacity jump on cooling below T_c along the critical isochore is [5, 6]

$$\frac{\Delta C_V(T_c)}{Nk_B} = 6 - \frac{3}{2} = \frac{9}{2}. \quad (7.23)$$

The first two terms in the second of equations (7.22*a*) are the same as given by Stanley [6].

Reichl has obtained a general expression for the heat capacity of a fluid at constant volume along an isochoric thermodynamic path within the gas–liquid coexistence region [5]. Her expression in our notation is

$$\frac{C_V}{Nk_B} = \frac{3}{2} - \frac{T}{N}\left[\frac{N_g}{N}\left(\frac{\partial p}{\partial V}\right)_T\Bigg|_{V=V_g}\left(\frac{\partial V_g}{\partial T}\right)^2 + \frac{N_l}{N}\left(\frac{\partial p}{\partial V}\right)_T\Bigg|_{V=V_l}\left(\frac{\partial V_l}{\partial T}\right)^2\right], \quad (7.24)$$

where p is the pressure of the pure fluid at volume V, which is obtained from the equation of state for the fluid.

In terms of our reduced variables for the vdW fluid, equation (7.24) becomes

$$\frac{C_V(y, \hat{V}_0)}{Nk_B} = \frac{3}{2} - \frac{3\hat{\tau}}{8}\left[\frac{N_g}{N}\left(\frac{\partial \hat{p}}{\partial \hat{V}}\right)_{\hat{\tau}}\Bigg|_{\hat{V}=\hat{V}_g}\left(\frac{\partial \hat{V}_g}{\partial \hat{\tau}}\right)^2 + \frac{N_l}{N}\left(\frac{\partial \hat{p}}{\partial \hat{V}}\right)_{\hat{\tau}}\Bigg|_{\hat{V}=\hat{V}_l}\left(\frac{\partial \hat{V}_l}{\partial \hat{\tau}}\right)^2\right]. \quad (7.25a)$$

The vdW equation of state equation (5.5) yields

$$\left(\frac{\partial \hat{p}}{\partial \hat{V}}\right)_{\hat{\tau}}\Bigg|_{\hat{V}=\hat{V}_{g,l}} = \frac{6}{\hat{V}_{g,l}^3} - \frac{24\hat{\tau}}{\left(3\hat{V}_{g,l} - 1\right)^2}. \quad (7.25b)$$

The molar fractions N_g/N and N_l/N are given in terms of $\hat{V}_g(y)$, $\hat{V}_l(y)$ and $\hat{V}_0$ in equations (6.10), and the temperature derivatives in equation (7.25*a*) are handled as described in equation (7.21*c*). From equations (7.25) we obtained a parametric solution for $C_V/(Nk_B)$ versus y. We checked the consistency of our solution from Reichl's treatment with our previous solution in equation (C.1) of the appendix by calculating $C_V(y)/(Nk_B)$ for $\hat{V} = 1$ from equations (7.25) and (7.3), and find that the two solutions are identical.

References

[1] Lekner J 1982 Parametric solution of the van der Waals liquid–vapor coexistence curve *Am. J. Phys.* **50** 161

[2] Berberan-Santos M N, Bodunov E N and Pogliani L 2008 The van der Waals equation: analytical and approximate solutions *J. Math. Chem.* **43** 1437

[3] Guggenheim E A 1945 The principle of corresponding states *J. Chem. Phys.* **13** 253

[4] Johnston D C, McQueeney R J, Lake B, Honecker A, Zhitomirsky M E, Nath R, Furukawa Y, Antropov V P and Singh Y 2011 Magnetic exchange interctions in $BaMn_2As_2$: a case study of the J_1-J_2-J_c Heisenberg model *Phys. Rev.* B **84** 094445

[5] Reichl L E 2009 *A Modern Course in Statistical Physics* 3rd edn (Weinheim: Wiley-VCH)

[6] Stanley H E 1971 *Introduction to Phase Transitions and Critical Phenomena* (New York: Oxford Science)

Chapter 8

Static critical exponents

We introduce the following notations which are useful when considering the approach to the critical point:

$$\tau_0 \equiv \hat{\tau} - 1, \qquad t_0 \equiv -\tau_0 = 1 - \hat{\tau}, \qquad v_0 \equiv \hat{V} - 1,$$
$$n_0 \equiv \hat{n} - 1, \qquad p_0 \equiv \hat{p} - 1, \qquad \mu_0 \equiv \frac{\mu - \mu_c}{\tau_c}, \tag{8.1}$$

where μ_c is the chemical potential at the critical point. The notation t_0 was previously introduced in equations (7.5) in the context of the coexistence curve.

The asymptotic critical exponents relate the changes in a property of a system to an infinitesimal deviation of a state variable from the critical point. The definitions of some static critical exponents relevant to the thermodynamics of the vdW fluid are given in table 8.1. Experimental data (see, e.g. [1–3]) indicate that the liquid–gas transition belongs to the universality class of the three-dimensional (3D) Ising model, which is a 3D model with short-range interactions and a scalar order parameter [3]. The theoretical values for the critical exponents α, β, γ and δ for this model are given in table 8.1 [3], where the value of δ is obtained from the scaling law $\beta\delta = \beta + \gamma$ [4]. Also shown in table 8.1 are well-known critical exponents for the mean-field vdW fluid [4–7]. One sees that the vdW exponents are in poor agreement with the 3D Ising model predictions and therefore also in poor agreement with the corresponding experimental values.

In the following we derive the vdW exponents for the static critical behaviors together with the corresponding amplitudes expressed in our dimensionless reduced forms. Some of these exponents and amplitudes are needed in chapter 10 for comparison with our numerical calculations at temperatures near T_c.

doi:10.1088/978-1-627-05532-1ch8

Table 8.1. Static critical exponents for liquid–gas phase transitions. The parameter $\tau_0 \equiv \hat{\tau} - 1 = \frac{T}{T_c} - 1$ measures the fractional deviation of the temperature from the critical temperature T_c. The unprimed critical exponents are for $\tau_0 > 1$ and the primed ones for $\tau_0 < 1$. The prefactors of the powers of τ_0 are the critical amplitudes. The critical exponent α is for the heat capacity at constant volume, β is for the liquid–gas number density difference (order parameter) $\hat{n}_l - \hat{n}_g = (\frac{1}{\hat{V}_l} - \frac{1}{\hat{V}_g})$ on traversing the liquid–gas coexistence curve such as in figure 6.7, γ and γ_p are for the isothermal compressibility and δ is for the critical p–V isotherm which is the p–V isotherm that passes through the critical point $\hat{\tau} = \hat{p} = \hat{V} = 1$. The experimental critical exponents of fluids are described well by the exponents for the 3D Ising model as shown [3]. The classical mean-field critical exponents and amplitudes for the vdW fluid are in the last two columns. The definitions of the critical exponents except for γ_p and γ_p' are from [6].

Exponent	Definition	Thermodynamic path	3D Ising model exponent	vdW exponent	vdW amplitude
α	$C_V/(Nk_B) = a\,\tau_0^{-\alpha}$	$\tau_0 > 0;\ \hat{V} = 1$	0.110(3)	$\alpha = 0$	$a = \frac{3}{2}$
α'	$C_V/(Nk_B) = a'(-\tau_0)^{-\alpha'}$	$\tau_0 < 0;\ \hat{V} = 1$		$\alpha' = 0$	$a' = 6$
β	$\hat{n}_l - \hat{n}_g = b(-\tau_0)^{\beta}$	$\tau_0 < 0;\ \hat{p}$–$\hat{\tau}$ coexistence curve	0.326(2)	$\beta = \frac{1}{2}$	$b = 4$
γ	$\kappa_T p_c = g\,\tau_0^{-\gamma}$	$\tau_0 > 0;\ \hat{V} = 1$	1.239(2)	$\gamma = 1$	$g = \frac{1}{6}$
γ'	$\kappa_T p_c = g'(-\tau_0)^{-\gamma'}$	$\tau_0 < 0;\ \hat{p}$–$\hat{V}$ coexistence curves		$\gamma' = 1$	$g' = \frac{1}{12}$
γ_p	$\kappa_T p_c = g_p\,\tau_0^{-\gamma_p}$	$\tau_0 > 0;\ \hat{p} = 1$		$\gamma_p = \frac{2}{3}$	$g_p = \frac{1}{3^{1/3}6}$
γ_p'	$\kappa_T p_c = g_p'(-\tau_0)^{-\gamma_p'}$	$\tau_0 < 0;\ \hat{p} = 1$		$\gamma_p' = \frac{2}{3}$	$g_p' = \frac{1}{3^{1/3}6}$
δ	$p_0 = d\lvert n_0\rvert^{\delta}\,\mathrm{sgn}(n_0)$	$\tau_0 = 0;\ p_0,\ n_0 \neq 0$	4.80 (derived)	$\delta = 3$	$d = \frac{3}{2}$

8.1 Heat capacity at constant volume

Here we consider the thermodynamic path along the critical isochore that was treated in section 7.5. For $T \to T_c^+$, equation (4.7) gives

$$\frac{C_V(\hat{\tau} \to 1^+)}{Nk_B} = \frac{3}{2} \equiv a\,(\tau_0)^{-\alpha}, \tag{8.2}$$

yielding the critical exponent and amplitude

$$\alpha = 0 \qquad a = \frac{3}{2}. \tag{8.3}$$

Similarly, from the second of equations (7.22a) one obtains

$$\frac{C_V(\hat{\tau} \to 1^-)}{Nk_B} = 6 \equiv a'(t_0)^{-\alpha'}, \tag{8.4}$$

yielding

$$\alpha' = 0 \qquad a' = 6. \tag{8.5}$$

8.2 Pressure versus volume isotherm at the critical temperature

For small deviations $p_0 = \hat{p} - 1$ and $n_0 = \hat{n} - 1$ of $\hat{p}$ and $\hat{n}$ from their critical values of unity and setting $\hat{\tau} = 1$, a Taylor expansion of p_0 to lowest order in n_0 obtained from equation (5.9) gives

$$p_0 = \frac{3}{2}n_0^3. \tag{8.6}$$

A comparison of this result with the corresponding expression in table 8.1 yields the critical exponent δ and amplitude d as

$$\delta = 3 \qquad d = \frac{3}{2} \qquad (\hat{n} \to 1^{+,-}). \tag{8.7}$$

Thus the critical exponent and amplitude are the same on both sides of the critical point $\hat{n} < 1$ and $\hat{n} > 1$. To determine the temperature region over which the critical behavior approximately describes the critical isotherm, shown in figure 8.1 is a log–log plot of $|p_0| \equiv |\hat{p} - 1|$ versus $|n_0| \equiv |\hat{n} - 1|$. The data are seen to follow the predicted asymptotic critical behavior $p_0 = d\, n_0^\delta$ with amplitude $d = 3/2$ and exponent $\delta = 3$ for $0.9 \lesssim \hat{n} \lesssim 1.1$. This region with $\hat{p} \sim 1 \pm 0.001$ appears horizontal on the scale of the plot with $\hat{\tau} = 1$ in figure 5.3.

8.3 Critical chemical potential isotherm versus number density

From equation (5.37c), there is no law of corresponding states for the behavior of the chemical potential of a vdW fluid near the critical point unless one only considers

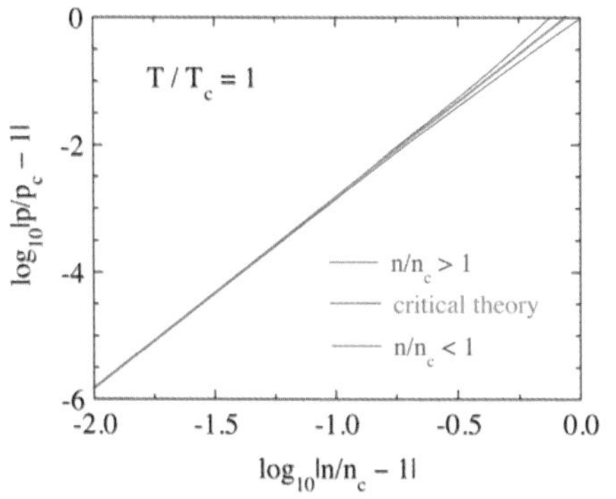

Figure 8.1. Log–log plot of $|\hat{p} - 1|$ versus $|\hat{n} - 1|$ for the critical $\hat{p}$ versus $\hat{n}$ isotherm at $\hat{\tau} = 1$ in figure 5.3 obtained from equation (5.9). On the far right, the top red solid curve is for $n/n_c > 1$ and the bottom blue solid curve is for $n/n_c < 1$. The predicted asymptotic critical behavior $|\hat{p} - 1| = d|\hat{n} - 1|^\delta$ with $d = 3/2$ and $\delta = 3$ in equation (8.7) is shown by the green line.

processes on the critical isotherm for which $\hat{\tau} = 1$. The value of the chemical potential at the critical point is

$$\frac{\mu_c}{\tau_c} = -\left[\ln(2X) + \frac{7}{4}\right]. \tag{8.8}$$

Expanding equation (5.37*c*) with $\hat{\tau} = 1$ in a Taylor series to the lowest three orders in $n_0 \equiv \hat{n} - 1$ gives

$$\mu_0 \equiv \frac{\mu - \mu_c}{\tau_c} = \frac{9n_0^3}{16} - \frac{9n_0^4}{64} + \frac{81n_0^5}{320}. \tag{8.9}$$

Comparing the first term of this expression with the critical behavior of the pressure in equation (8.6), one obtains

$$\mu_0 = \frac{3p_0}{8} = \frac{9n_0^3}{16} \qquad \left(\hat{n} \to 1^{+,-}\right). \tag{8.10}$$

Thus the critical exponent is the same as $\delta = 3$ in table 8.1 for the critical p–V isotherm but the amplitude is smaller than $d = \frac{3}{2}$ by a factor of 3/8.

8.4 Liquid–gas transition order parameter

We now determine the critical behavior of the difference in density between the liquid and gas phases on the coexistence curve, which is the order parameter for the liquid–gas transition. Equation (7.7*k*) gives the asymptotic critical behavior as

$$\Delta\hat{n} \equiv \hat{n}_l - \hat{n}_g = 4\sqrt{-\tau_0}. \tag{8.11}$$

Comparison of this expression with the definitions in table 8.1 gives the critical exponent and amplitude of the order parameter of the transition as

$$\beta = \frac{1}{2} \qquad b = 4 \qquad \left(\hat{\tau} \to 1^{-}\right). \tag{8.12}$$

The exponent is typical of mean-field theories of second-order phase transitions. The transition at the critical point is second order because the latent heat goes to zero at the critical point (figure 7.5(*c*) above). Since the order parameter is necessarily zero for $\hat{\tau} > 1$, one does not define a critical exponent and amplitude for this path of approach to the critical point.

Figure 8.2(*a*) shows an expanded plot of the data in figure 7.1(*b*) of the difference between the densities of the coexisting liquid and gas phases versus temperature. One sees a sharp downturn as T approaches T_c. In figure 8.2(*b*) is plotted $\log_{10}\Delta\hat{n}$ versus $\log_{10}(1 - \hat{\tau})$. For $1 - \hat{\tau} \lesssim 10^{-3}$, we obtain $\Delta\hat{n} = 3.999(1 - \hat{\tau})^{0.4999}$, consistent with

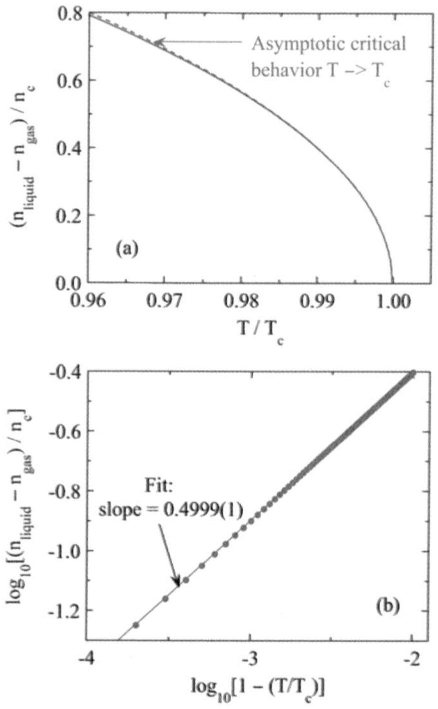

Figure 8.2. (*a*) Expanded plot from figure 7.1(*b*) of the difference $\Delta\hat{n} = \hat{n}_l - \hat{n}_g$ between the reduced liquid and gas densities near the critical point $\hat{\tau} = T/T_c = 1$ versus reduced temperature $\hat{\tau}$ (red curve). $\Delta\hat{n}$ is the order parameter for the gas–liquid transition. (*b*) Logarithm to the base 10 of $\Delta\hat{n}$ versus the logarithm to the base 10 of the difference $1 - \hat{\tau}$. The fitted straight line for the data points on the lower left with $1 - \hat{\tau} < 10^{-3}$ is given by $\Delta\hat{n} = b(1 - \hat{\tau})^{\beta}$ with $b = 3.999$ and $\beta = 0.4999$, consistent with the amplitude $b = 4$ and exponent $\beta = 1/2$ predicted in equation (8.12). This asymptotic critical behavior is shown as the dashed blue curve in panel (a), where this behavior is seen to be followed fairly accurately for $\hat{\tau} \gtrsim 0.97$.

the critical exponent and amplitude in equations (8.12). The dashed blue curve in figure 8.2(*a*) is a plot of the critical behavior, which is seen to be followed fairly well for $\hat{\tau} \gtrsim 0.97$.

8.5 Isothermal compressibility

The critical behaviors of $\kappa_{\rm T}$ for $\hat{\tau} \to 1^{+,-}$ are obtained from equation (5.27*a*) according to

$$\frac{1}{\kappa_{\rm T} p_{\rm c}} = -\hat{V}\left(\frac{\partial \hat{p}}{\partial \hat{V}}\right)_{\hat{\tau}}\Bigg|_{\hat{V}\to 1,\ \hat{\tau}\to 1}. \tag{8.13}$$

From equation (5.5), one obtains

$$\left(\frac{\partial \hat{p}}{\partial \hat{V}}\right)_{\hat{\tau}} = -\frac{24\hat{\tau}}{\left(3\hat{V}-1\right)^2} + \frac{6}{\hat{V}^3}. \tag{8.14}$$

Writing this expression in terms of the expansion parameters in equations (8.1) and Taylor expanding to lowest orders gives

$$\left(\frac{\partial \hat{p}}{\partial \hat{V}}\right)_{\hat{\tau}} = -6\tau_0 - \frac{9}{2}v_0^2 \qquad (\hat{\tau} > 1), \tag{8.15a}$$

$$\left(\frac{\partial \hat{p}}{\partial \hat{V}}\right)_{\hat{\tau}} = 6t_0 - \frac{9}{2}v_0^2 \qquad (\hat{\tau} < 1). \tag{8.15b}$$

8.5.1 Approach to the critical point along the critical isochore from above the critical temperature

Setting $v_0 = 0$, equations (8.13) and (8.15*a*) immediately give

$$\kappa_{\rm T} p_{\rm c} = \frac{1}{6\tau_0}. \tag{8.16}$$

Then the definition of the critical behavior of $\kappa_{\rm T}$ in table 8.1 gives the critical exponent γ and amplitude g as

$$\gamma = 1 \qquad g = \frac{1}{6} \qquad \left(\hat{V} = 1,\ \hat{\tau} \to 1^{+}\right). \tag{8.17}$$

8.5.2 Approach to the critical point along either boundary of the gas–liquid coexistence region for temperatures below the critical temperature

Defining the isothermal compressibility at either the pure gas or pure liquid coexistence points G or C on the p–V isotherm in figure 6.2 has been used to define the critical behavior of $\kappa_{\rm T}$ for $\hat{\tau} < 1$. The value of $(\partial\hat{p}/\partial\hat{V})_{\hat{\tau}}$ is the slope of a p–V isotherm for $\hat{\tau} \to 1^{-}$ which can be specified for either of those points since only their magnitudes are relevant to equation (8.18) below. Referring to figure 6.2, the reduced value of the volume V_G is what we called $\hat{V}_l$ for the coexisting liquid phase in equations (7.3) and V_C corresponds to $\hat{V}_g$ for the coexisting gas phase. For either the liquid or gas phases, to lowest order in t_0 equations (7.7) give

$$v_{0\,l,g}^2 = 4\,t_0. \tag{8.18}$$

Substituting this value into equation (8.15*b*) gives

$$\left(\frac{\partial \hat{p}}{\partial \hat{V}}\right)_{\hat{\tau}} = -12t_0. \tag{8.19}$$

Then equation (8.13) becomes

$$\kappa_T p_c = \frac{1}{12t_0}, \tag{8.20}$$

so the critical exponent γ' and amplitude g' are

$$\gamma' = 1 \qquad g' = \frac{1}{12} \qquad (\hat{\tau} \to 1^-). \tag{8.21}$$

Comparing equations (8.17) and (8.21), the critical exponents are the same for $\hat{\tau} \to 1^+$ and $\hat{\tau} \to 1^-$ but the amplitudes are a factor of two different [6]. In the following section the critical exponents and amplitudes of κ_T are found to be different from the above values when the critical point is approached along the critical isobar.

8.6 Approach to the critical point along the critical isobar

In this section we consider the critical exponents and amplitudes of κ_T, α and C_p on approaching the critical point along the critical isobar, i.e. $\hat{p} = 1$. We need these to compare with corresponding numerical calculations in section 10.2 below. Setting $\hat{p} = 1$, the equation of state (5.4) becomes

$$\hat{\tau} = \left(1 + \frac{3}{\hat{V}^2}\right)\frac{3\hat{V} - 1}{8}. \tag{8.22}$$

The lowest-order Taylor series expansion of this equation in the variables τ_0 and v_0 in equations (8.1) gives

$$\tau_0 = \frac{3}{8}v_0^3,$$

so

$$v_0 = \frac{2|\tau_0|^{1/3}}{3^{1/3}}\,\mathrm{sgn}(\tau_0) \qquad (\hat{p} = 1), \tag{8.23}$$

where these expressions are valid for both signs of v_0 and τ_0 which are the same for the two quantities.

8.6.1 Isothermal compressibility

For $\hat{\tau} > 1$, substituting v_0^2 from equation (8.23) into (8.15) and using the definition in equation (8.13) gives, after a Taylor series expansion,

$$\kappa_{\rm T} p_{\rm c} = \frac{1}{3^{1/3}6} |\tau_0|^{-2/3} \qquad (\hat{p} = 1, \ \hat{\tau} \to 1^{+,-}). \tag{8.24a}$$

Thus the critical exponents and amplitudes as defined in table 8.1 are

$$\begin{aligned} \gamma_{\rm p} &= \gamma_{\rm p}' = \frac{2}{3}, \\ g_{\rm p} &= g_{\rm p}' = \frac{1}{3^{1/3}6} \approx 0.1156, \end{aligned} \tag{8.24b}$$

which are the same for $\hat{\tau} > 1$ and $\hat{\tau} < 1$ as discussed above (see also the critical isobar in the V–T plane in figure 6.6). The critical exponents $\gamma_{\rm p}$ and $\gamma_{\rm p}'$ and respective amplitudes g and g' are not commonly quoted.

8.6.2 Volume thermal expansion coefficient

From equation (5.30), near the critical point and on a path with $\hat{p} = 1$, one has

$$\left.\frac{\alpha\tau_{\rm c}}{k_{\rm B}}\right|_{\hat{p}=1} = \left.\left(\frac{\partial \hat{p}}{\partial \hat{\tau}}\right)_{\hat{V}}\right|_{\hat{V}=1} \kappa_{\rm T} p_{\rm c}. \tag{8.25}$$

From equation (5.5), the derivative is

$$\left.\left(\frac{\partial \hat{p}}{\partial \hat{\tau}}\right)_{\hat{V}}\right|_{\hat{V}=1} = 4. \tag{8.26}$$

Inserting this result and equation (8.24a) into (8.25) gives

$$\frac{\alpha\tau_{\rm c}}{k_{\rm B}} = 4\kappa_{\rm T} p_{\rm c} = \frac{2}{3^{4/3}} |\tau_0|^{-2/3} \approx 0.4622 |\tau_0|^{-2/3} \qquad (\hat{p} = 1). \tag{8.27}$$

Thus $\alpha\tau_{\rm c}/k_{\rm B}$ has the same critical exponent as $\kappa_{\rm T} p_{\rm c}$ but with an amplitude that is four times larger than for $\kappa_{\rm T} p_{\rm c}$.

8.6.3 Heat capacity at constant pressure

Inserting the above expressions for $\alpha\tau_{\rm c}/k_{\rm B}$ and $\kappa_{\rm T} p_{\rm c}$ near the critical point into the expression (5.33c) for $C_{\rm p}$ gives

$$\frac{C_{\rm p}}{Nk_{\rm B}} = \frac{3}{2} + \frac{1}{3^{1/3}} |\tau_0|^{-2/3} \approx \frac{3}{2} + 0.6934 |\tau_0|^{-2/3} \qquad (\hat{p} = 1). \tag{8.28}$$

When examining the critical part of C_p, one would remove the noncritical part 3/2 due to C_V from the right-hand side.

Some of the above critical exponents and amplitudes of the vdW fluid are listed in table 8.1. As seen above, the critical exponents and amplitudes for particular properties and thermodynamic paths are derivable from others [2, 7].

References

[1] Hocken R and Moldover M R 1976 Ising critical exponents in real fluids: an experiment *Phys. Rev. Lett.* **37** 29

[2] Levelt Sengers J M H, Morrison G and Chang R F 1983 Critical behavior in fluids and fluid mixtures *Fluid Phase Equilib.* **14** 19

[3] Sengers J V and Shanks J G 2009 Experimental critical-exponent values for fluids *J. Stat. Phys.* **137** 857

[4] Kadanoff L P, Götze W, Hamblen D, Hecht R, Lewis E A S, Palciauskas V V, Rayl M and Swift J 1967 Static phenomena near critical points: theory and experiment *Rev. Mod. Phys.* **39** 395

[5] Reichl L E 2009 *A Modern Course in Statistical Physics* 3rd edn (Weinheim: Wiley-VCH)

[6] Stanley H E 1971 *Introduction to Phase Transitions and Critical Phenomena* (New York: Oxford Science)

[7] Levelt Sengers J M H 1991 *Supercritical Fluid Technology: Reviews in Modern Theory and Applications* ed T J Bruno and J F Ely (Boca Raton FL: CRC Press) chapter 1 pp 1–56

David C Johnston

Chapter 9

Superheating and supercooling

It is well known that systems exhibiting first-order phase transitions can exhibit hysteresis in the transition temperature and therefore in other physical properties upon cooling and warming, where the transition temperature is lower on cooling (supercooling) and higher on warming (superheating) than the equilibrium transition temperature T_X. The vdW fluid can also exhibit these properties.

Shown in figure 9.1(a) is a plot of reduced volume $\hat{V}$ versus reduced temperature $\hat{\tau}$ at fixed pressure $\hat{p} = 0.3$ from figure 5.4 which is predicted from the vdW equation of state (5.4). Important points on the curve are labeled by numbers. Points 1 and 9 correspond to pure gas and liquid phases, respectively, and are in the same regions as points A and I in the p versus V isotherm in the top panel of figure 6.2. Points 3, 5 and 7 are at the reduced gas–liquid coexistence temperature $\hat{\tau}_X = T_X/T_c$ as in figure 6.2. Points 3 and 7 thus correspond to points C and G in figure 6.2, respectively. Points 4 and 6 are points of infinite slope of $\hat{V}$ versus $\hat{\tau}$. The volumes at points 4 and 6 do not correspond precisely with those at points D and F in figure 6.2 at the same pressure, contrary to what might have been expected. The curve 4–5–6 is not accessible to the vdW fluid because the thermal expansion coefficient is negative along this curve.

There is no mechanical constraint that prevents the system from following the path 1–2–3–4 in figure 9.1(a) on cooling, where point 4 overshoots the equilibrium phase transition temperature. When a liquid first nucleates as small droplets on cooling, the surface to volume ratio is large, and the surface tension (surface free energy) tends to prevent the liquid droplets from forming. This free energy is not included in the treatment of the bulk vdW fluid, and represents a potential energy barrier that must be overcome by density fluctuations (homogeneous nucleation) or by interactions of the fluid with a surface or impurities (heterogeneous nucleation) before a bulk phase transition can occur [1]. These mechanisms take time to nucleate sufficiently large liquid droplets and therefore rapid cooling promotes this so-called supercooling. The minimum possible reduced supercooling temperature $\hat{\tau}_{SC} = T_{SC}/T_c$ occurs at point 4 in figure 9.1(a), resulting in a supercooling curve given by the

doi:10.1088/978-1-627-05532-1ch9

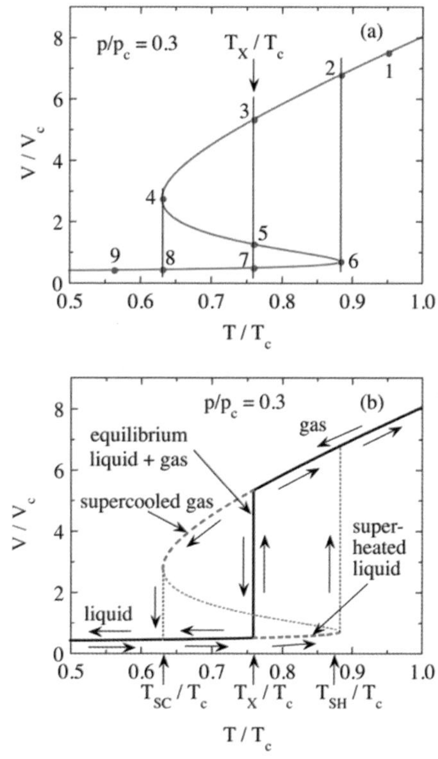

Figure 9.1. (*a*) Reduced volume $\hat{V} = V/V_c$ versus reduced temperature $\hat{\tau} = T/T_c$ at constant reduced pressure $\hat{p} = p/p_c = 0.3$ from figure 5.4. Significant points on the curve described in the text are labeled by numbers. The reduced temperature $\hat{\tau}_X = T_X/T_c$ is the equilibrium first-order phase transition temperature. (*b*) The equilibrium behavior of $\hat{V}$ versus $\hat{\tau}$ is shown by the heavy black vertical line, which denotes the region of liquid–gas coexistence as in figure 6.6. The paths of supercooled gas with minimum reduced temperature $\hat{\tau}_{SC} = T_{SC}/T_c$ and superheated liquid with maximum reduced temperature $\hat{\tau}_{SH} = T_{SH}/T_c$ are shown by dashed blue and red lines, respectively. The dotted green curve with negative slope is not accessible to the vdW fluid. The vertical blue and red dotted lines are not equilibrium mixtures of gas and liquid; these lines represent an irreversible spontaneous decrease and increase in volume, respectively.

dashed blue curve in figure 9.1(*b*). Similarly, superheating can occur with a maximum reduced temperature $\hat{\tau}_{SH} = T_{SH}/T_c$ at point 6 in figure 9.1(*a*), resulting in a superheating curve given by the dashed red curve in figure 9.1(*b*). The vertical dotted blue and red lines in figure 9.1(*b*) represent nonequilibrium irreversible transitions from supercooled gas to liquid and from superheated liquid to gas, respectively. The latter can be dangerous because this transition can occur rapidly, resulting in explosive spattering of the liquid as it transforms into gas with a much larger volume. This effect in superheated water is discussed in [2]. The dashed supercooling and superheating curves in figure 9.1(*b*) are included in the T versus V phase diagram in figure 6.4(*b*).

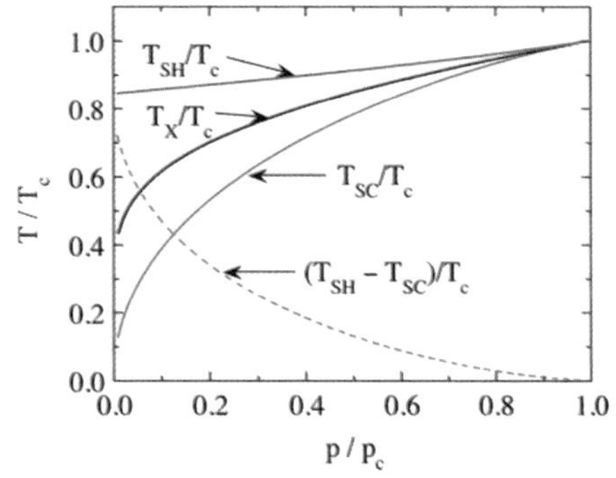

Figure 9.2. Reduced equilibrium phase transition temperature $\hat{\tau}_X = T_X/T_c$, maximum reduced superheating temperature $\hat{\tau}_{SH} = T_{SH}/T_c$, minimum reduced supercooling temperature $\hat{\tau}_{SC} = T_{SC}/T_c$ and the difference $\hat{\tau}_{SH} - \hat{\tau}_{SC}$ versus reduced pressure $\hat{p} = p/p_c$ from $\hat{p} = 0.01$ to 1.

The reduced volumes $\hat{V}_4$ and $\hat{V}_6$ in figure 9.1(*a*) are calculated for a given pressure $\hat{p}$ from the equation of state (5.4) as the volumes at which $d\hat{\tau}/d\hat{V} = 0$ (and $d\hat{V}/d\hat{\tau} = \infty$). Then the reduced temperatures $\hat{\tau}_{SC} = \hat{\tau}_4$ and $\hat{\tau}_{SH} = \hat{\tau}_6$ are determined from these volumes and the given $\hat{p}$ using equation (5.4). The equilibrium first-order transition temperature $\hat{\tau}_X$ is calculated by first finding the volumes $\hat{V}_3$ and $\hat{V}_7$ at which the chemical potentials in equation (5.33) are equal, where one also requires that $\hat{\tau}_3 = \hat{\tau}_7$ without explicitly calculating their values. Once these volumes are determined, the value of $\hat{\tau}_X = \hat{\tau}_3 = \hat{\tau}_7$ is determined from equation (5.4). Plots of $\hat{\tau}_X$, $\hat{\tau}_{SH}$, $\hat{\tau}_{SC}$ and $\hat{\tau}_{SH} - \hat{\tau}_{SC}$ are shown versus $\hat{p}$ from $\hat{p} = 0.01$ to 1 in figure 9.2. One sees that $\hat{\tau}_X$ is roughly midway between $\hat{\tau}_{SH}$ and $\hat{\tau}_{SC}$ over the whole pressure range, with $\hat{\tau}_{SH} - \hat{\tau}_{SC}$ decreasing monotonically with increasing temperature and going to zero at $\hat{p} = 1$ as expected.

A list of values of $\hat{\tau}_{SC}$, $\hat{\tau}_X$, $\hat{\tau}_{SH}$, $\hat{V}_7$, $\hat{V}_3$, $\hat{V}_5$, $\hat{V}_8$, $\hat{V}_4$, $\hat{V}_6$ and $\hat{V}_2$ versus $\hat{p}$ is given in table A.4 in appendix A.

References

[1] Kittel C and Kroemer H 1980 *Thermal Physics* 2nd edn (New York: Freeman)

[2] Imre A R, Baranyai A, Deiters U K, Kiss P T, Kraska T and Quiñones Cisneros S E 2013 Estimation of the thermodynamic limit of overheating for bulk water from interfacial properties *Int. J. Thermophys.* **34** 2053 and references therein

David C Johnston

Chapter 10

Additional numerical calculations of thermodynamic properties

10.1 Isotherms versus density in the supercritical temperature region

The thermodynamic properties of the vdW fluid can be calculated numerically from analytic formulas derived from the vdW equation of state. Here we calculate isotherms of several thermodynamic properties versus the number density in the supercritical region $T \geqslant T_c$. The required expressions are given above for κ_T, α and C_p in equations (5.28*b*), (5.31*b*) and (5.34*b*), respectively. Isotherms of κ_T, α and C_p are shown in figures 10.1(*b*), 10.1(*a*) and 10.2(*b*), respectively. Each of these quantities shows a critical divergence at $\hat{\tau} \to 1$ and $\hat{n} \to 1$. The divergences become rounded peaks for $\hat{\tau} > 1$ and the amplitudes of the peaks decrease with increasing $\hat{\tau}$, i.e. as the distance from the critical point increases. The peaks for all three properties are seen to strongly diminish by $\hat{\tau} = 1.3$.

Isotherms of the number density fluctuation versus density are also of interest [1]. The general expression for the density fluctuation amplitude is [2]

$$N_{\text{fluct}} \equiv \frac{\langle [N - \langle N \rangle]^2 \rangle}{\langle N \rangle} = \frac{\langle N \rangle k_B T}{V} \kappa_T, \tag{10.1a}$$

where $\langle \cdots \rangle$ is the thermal average of the enclosed quantity and $\langle N \rangle = N$ for a macroscopic value of N. In terms of our reduced variables in equations (5.3) and using equation (4.4*c*), one obtains

$$N_{\text{fluct}} = \frac{8\hat{\tau}\hat{n}}{3} \kappa_T p_c. \tag{10.1b}$$

Substituting the expression for $\kappa_T p_c$ in equation (5.28*b*) into equation (10.1*b*) gives [1]

$$N_{\text{fluct}} = \frac{4\hat{\tau}(3 - \hat{n})^2/9}{4\hat{\tau} - \hat{n}(3 - \hat{n})^2}. \tag{10.1c}$$

doi:10.1088/978-1-627-05532-1ch10

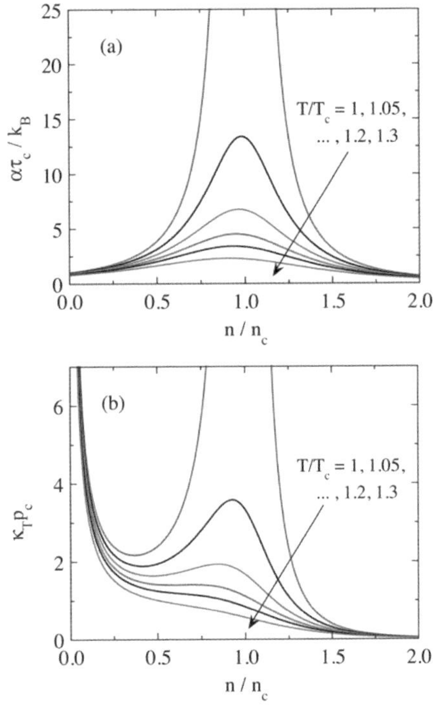

Figure 10.1. Isotherms of (*a*) the reduced thermal expansion coefficient $\alpha\tau_c/k_B$ and (*b*) reduced isothermal compressibility $\kappa_T p_c$ versus reduced number density $\hat{n} = n/n_c$ at the indicated reduced temperatures $\hat{\tau} = T/T_c$.

Isotherms of N_{fluct} versus $\hat{n}$ are shown in figure 10.2(*a*), where the overall characteristics are similar to those of the α, κ_T and C_p isotherms discussed above.

In general, the density at which the peak occurs in the above thermodynamic functions depends on the temperature. We define a 'ridge' or 'pseudocritical curve' as the locus of such peak positions in the temperature–density plane. Analytic expressions for the ridges are obtained by setting the first derivative of equations (5.28*b*), (5.31*b*), (5.34*b*) and (10.1*c*) for each of the functions with respect to density equal to zero, yielding [1, 3]

$$\hat{\tau}_{\alpha\,\text{ridge}} = \frac{1}{4}\left(3 - 2\hat{n}\right)\left(3 - \hat{n}\right)^2, \tag{10.2a}$$

$$\hat{\tau}_{\kappa_T\,\text{ridge}} = \frac{\hat{n}\left(3 - \hat{n}\right)^3}{2\left(3 + \hat{n}\right)}, \tag{10.2b}$$

$$\hat{\tau}_{C_p\,\text{ridge}} = 1 \qquad \left(\hat{n} = 1\right), \tag{10.2c}$$

$$\hat{\tau}_{N_{\text{fluct}}\,\text{ridge}} = \frac{1}{8}\left(3 - \hat{n}\right)^3. \tag{10.2d}$$

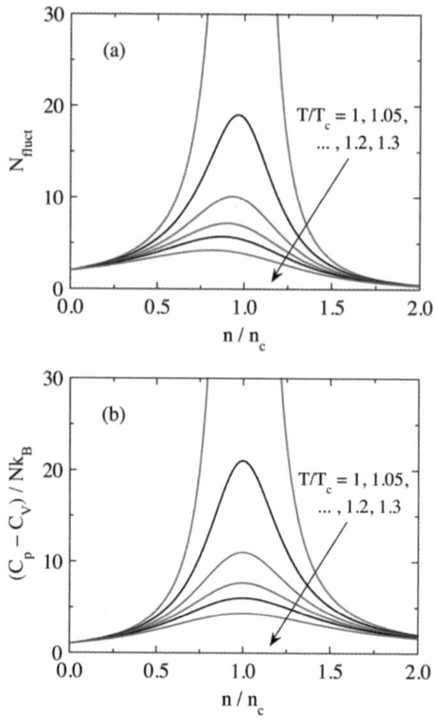

Figure 10.2. Isotherms of (*a*) the reduced density fluctuation amplitude $N_{\rm fluct}$ and (*b*) the reduced heat capacity at constant pressure minus the heat capacity at constant volume $(C_{\rm P} - C_{\rm V})/Nk_{\rm B}$ versus reduced number density $\hat{n} = n/n_{\rm c}$ at the indicated reduced temperatures $\hat{\tau} = T/T_{\rm c}$.

One sees that all four expressions yield $\hat{\tau}_{\rm ridge}(\hat{n} \to 1) = 1$, as expected since all four ridges originate at the critical point with $\hat{n} = \hat{\tau} = 1$.

Now we determine the limits on $\hat{n}$ for the four ridges in the supercritical temperature range $\hat{\tau}_{\rm ridge} \geqslant 1$. Equations (10.2) give

$$\alpha\ \text{ridge}\quad 0 < \hat{n} \leqslant 1, \tag{10.3a}$$

$$\kappa_{\rm T}\ \text{ridge}\quad 0.37108 \leqslant \hat{n} \leqslant 1, \tag{10.3b}$$

$$C_{\rm p}\ \text{ridge}\quad \hat{n} = 1 \qquad (\hat{\tau} \geqslant 1), \tag{10.3c}$$

$$N_{\rm fluct}\ \text{ridge}\quad 0 < \hat{n} \leqslant 1. \tag{10.3d}$$

For the two properties α and $N_{\rm fluct}$ that allow $\hat{n} = 0$ on the ridge, from equations (10.2) one obtains [3]

$$\hat{\tau}_{\alpha\,\rm ridge}(\hat{n} = 0) = \frac{27}{4}, \tag{10.4a}$$

$$\hat{\tau}_{N_{\text{fluct}}\,\text{ridge}}(\hat{n} = 0) = \frac{27}{8}. \tag{10.4b}$$

Shown in figure 10.3 are plots of the ridges near the critical point in the temperature–density (T–n) plane of α, κ_T, C_p and N_{fluct} obtained using equations (10.2), subject to the constraints in equations (10.3) that $\hat{\tau} \geqslant 1$. One sees that the ridges show no universal behavior even for $\hat{\tau} \to 1$. Furthermore, the ridge for κ_T shows a nonmonotonic dependence on n. Also shown in figure 10.3 for the temperature range $\hat{\tau} \leqslant 1$ are the T versus n plots for the gas–liquid coexistence curves $\hat{n}_{\text{gas}}$ and $\hat{n}_{\text{liquid}}$ from figure 7.2, together with their average $\hat{n}_{\text{ave}}$ from figure 7.4. The ridges for N_{fluct} and α at $\hat{\tau} \geqslant 1$ are similar to the extrapolation to $\hat{\tau} \geqslant 1$ of T versus $\hat{n}_{\text{ave}}$ at $\hat{\tau} \leqslant 1$.

Using the equation of state (5.9), one can write the ridge equations of $\hat{\tau}_{\text{ridge}}$ versus $\hat{n}$ in equations (10.2) instead as $\hat{p}_{\text{ridge}}$ versus $\hat{n}$ or versus $\hat{\tau}$ according to

$$\hat{p}_{\alpha\,\text{ridge}} = \hat{n}\left(18 - 21\hat{n} + 4\hat{n}^2\right), \tag{10.5a}$$

$$\hat{p}_{\kappa_T\,\text{ridge}} = \frac{\hat{n}^2\left(27 - 27\hat{n} + 4\hat{n}^2\right)}{3 + \hat{n}}, \tag{10.5b}$$

$$\hat{p}_{C_p\,\text{ridge}} = 4\hat{\tau} - 3, \tag{10.5c}$$

$$\hat{p}_{N_{\text{fluct}}\,\text{ridge}} = -27 + 36\hat{\tau}^{1/3} - 8\hat{\tau}. \tag{10.5d}$$

Setting $\hat{n} = \hat{\tau} = 1$ in these expressions yields $\hat{p}_{\text{ridge}} = 1$, as expected.

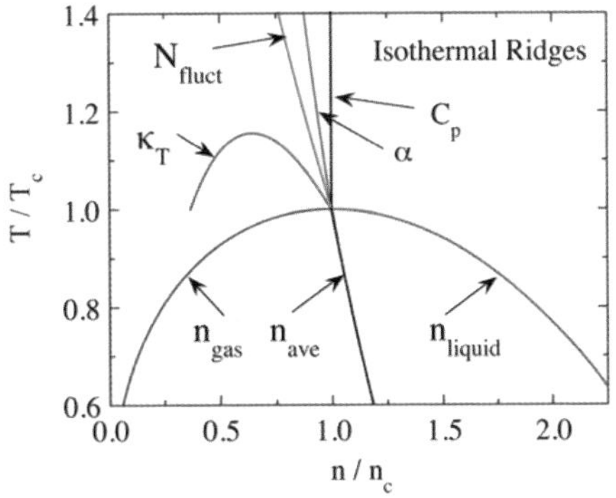

Figure 10.3. Isothermal ridges near the critical point for $\hat{\tau} \geqslant 1$ in the reduced temperature–density $\hat{\tau}$–$\hat{n}$ plane for the density fluctuation amplitude N_{fluct}, isothermal compressiblity κ_T, heat capacity at constant pressure C_p and thermal expansion coefficient α that occur in the data in figures 10.1 and 10.2. The analytic expressions for the ridges are given in equations (10.2) subject to the constraints in equations (10.3). Also shown are the gas–liquid coexistence curves $\hat{n}_{\text{gas}}$ and $\hat{n}_{\text{liquid}}$ and their average $\hat{n}_{\text{ave}}$ versus $\hat{n}$ for $\hat{\tau} \leqslant 1$ from figures 7.2 and 7.4.

If one now requires that $\hat{p} \geqslant 1$ in addition to $\hat{\tau} \geqslant 1$ in the supercritical regime, the limits of the ranges of $\hat{\tau}$ and $\hat{n}$ over which the ridges occur become

$$\alpha \text{ ridge} \qquad \begin{aligned} &1 \geqslant \hat{n} \geqslant \frac{17 - \sqrt{273}}{8} \approx 0.05966 \\ &1 \leqslant \hat{\tau} \leqslant \frac{7\left(79 + 9\sqrt{273}\right)}{256} \approx 6.2263, \end{aligned} \tag{10.6a}$$

$$\kappa_{\mathrm{T}} \text{ ridge} \qquad \begin{aligned} &1 \geqslant \hat{n} \geqslant 0.48411 \\ &1 \leqslant \hat{\tau} \leqslant 1.1064, \end{aligned} \tag{10.6b}$$

$$C_{\mathrm{p}} \text{ ridge} \qquad \hat{n} = 1, \quad \hat{\tau} \geqslant 1, \tag{10.6c}$$

$$N_{\mathrm{fluct}} \text{ ridge} \qquad \begin{aligned} &1 \leqslant \hat{\tau} \leqslant \frac{9\sqrt{15} - 23}{4} \approx 2.964 \\ &1 \geqslant \hat{n} \geqslant 4 - \sqrt{15} \approx 0.1270. \end{aligned} \tag{10.6d}$$

These constraints further limit the ranges (not shown) of three of the ridges in figure 10.3, where only the constraint $\hat{\tau} \geqslant 1$ was imposed.

10.2 Isobars versus temperature

The reduced number density $\hat{n} \equiv n/n_{\mathrm{c}} = 1/\hat{V}$ is plotted versus $\hat{\tau} \equiv T/T_{\mathrm{c}}$ in figure 10.4 for isobars with $\hat{p} \equiv p/p_{\mathrm{c}} = 0.3$ to 1.5 in 0.1 increments from equation (5.9). From figure 7.1(*a*), the maximum range of $\hat{n}$ is $0 \leqslant \hat{n} \leqslant 3$. As discussed above, the discontinuity in $\hat{n}$ on cooling below T_{c} as seen in figure 10.4(*b*) is the difference between the liquid and gas phase densities at the respective temperature and is the order parameter for the first-order liquid–gas phase transition. Therefore the height of the vertical discontinuity versus temperature in figure 10.4(*b*) is the order parameter versus temperature plotted previously in figure 7.1(*b*).

In the following sections, isobars of α, κ_{T} and C_{p} versus temperature in the pressure ranges $p \geqslant p_{\mathrm{c}}$ and $p \leqslant p_{\mathrm{c}}$ are presented and discussed separately. The supercritical regime with $\hat{\tau} \geqslant 1$ and $\hat{p} \geqslant 1$ is of special interest with respect to pseudocritical lines and the Widom line for $\hat{\tau} \to 1^{+}$ as discussed in section 10.2.2.

10.2.1 Results for pressures above the critical pressure

The entropy relative to that at the critical point $\Delta S/Nk_{\mathrm{B}}$ versus reduced temperature $\hat{\tau} = T/T_{\mathrm{c}}$ calculated using equations (5.5) and (5.26) for isobars with $\hat{p} = p/p_{\mathrm{c}} \geqslant 1$ with $\hat{V}$ as an implicit parameter is shown in figure 10.5(*a*). As the pressure decreases towards the critical pressure $\hat{p} = 1$, an inflection point develops in ΔS versus T with

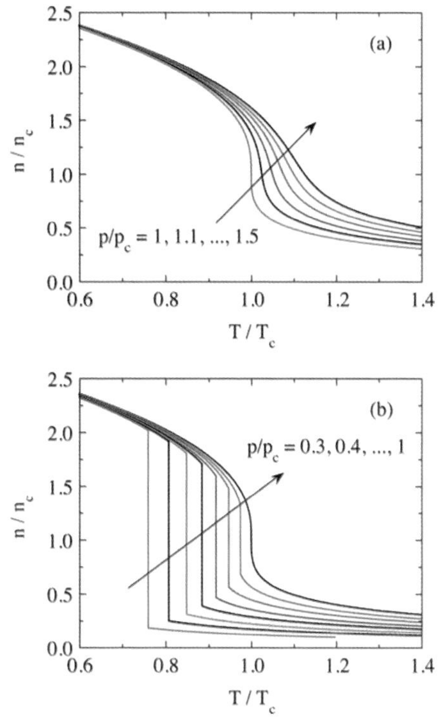

Figure 10.4. Equilibrium isobars of the reduced number density $\hat{n} = n/n_c$ versus reduced temperature $\hat{\tau} = T/T_c$ calculated using equation (5.9) for (*a*) $\hat{p} = p/p_c \geqslant 1$ and (*b*) $\hat{p} \leqslant 1$, where the two-phase liquid–gas regions are included (vertical lines).

a slope that increases to ∞ at $\hat{\tau} = 1$, signaling entrance into the phase-separated temperature range $\hat{\tau} < 1$ with decreasing pressure. The development of an infinite slope in ΔS versus T with decreasing pressure results in the onset of a divergence in C_p at the critical point discussed below. Similar behaviors are found for the internal energy and enthalpy using equations (5.5), (5.18) and (5.24) as shown in figures 10.6(*a*) and 10.7(*a*), respectively.

The thermal expansion coefficient $\alpha\tau_c/k_B$ versus T/T_c calculated from equations (5.5) and (5.31*a*) is plotted in figure 10.8(*a*) for $p/p_c = 1$ to 1.5 in 0.1 increments. The data show divergent behavior for $\hat{p} = 1$ at $T \to T_c$ which is found in figure 10.8(*b*) to be given by

$$\alpha\tau_c/k_B = 0.462(16)\left|\frac{T}{T_c} - 1\right|^{-0.667(2)}$$

for both $T \to T_c^{\pm}$, where the exponent and amplitude are equal to the analytical values of $-2/3$ and $2/3^{4/3}$ in equation (8.27) to within the error bars.

The isothermal compressibility $\kappa_T p_c$ versus T/T_c calculated from $\alpha\tau_c/k_B$ and equations (5.5) and (5.32) is plotted in figure 10.8(*c*) for $p/p_c = 1$ to 1.5 in 0.1 increments.

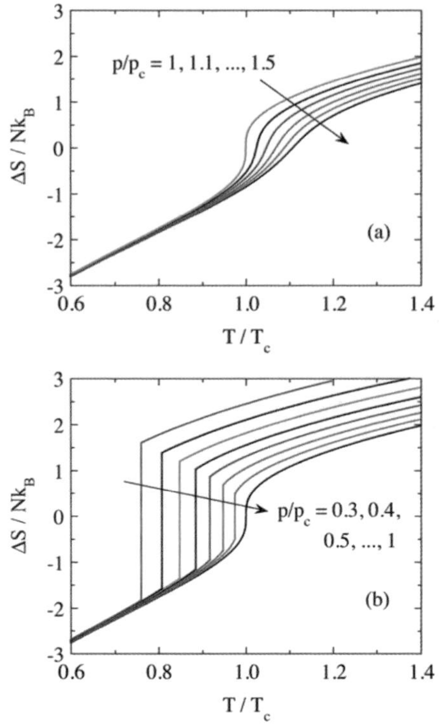

Figure 10.5. Equilibrium isobars of the difference in entropy $\Delta S/Nk_{\rm B}$ from that at the critical point with $\hat{\tau} = 1$ versus reduced temperature $\hat{\tau} = T/T_{\rm c}$ for (*a*) $\hat{p} = p/p_{\rm c} \geqslant 1$ and (*b*) $\hat{p} \leqslant 1$, where the liquid–gas coexistence regions are indicated by vertical straight lines.

The data again show divergent behavior for $\hat{p} = 1$ at $T \to T_{\rm c}$ which is found in figure 10.8(*d*) to be given by

$$\kappa_{\rm T} p_{\rm c} - 1 = 0.114(22) \left| \frac{T}{T_{\rm c}} - 1 \right|^{-0.668(11)}$$

for both $T \to T_{\rm c}^{\pm}$, where the exponent and amplitude are equal to the analytical values of $-2/3$ and $1/(3^{1/3}6)$ in equations (8.24*b*) to within the error bars. The non-critical background compressibility of the ideal gas $\kappa_{\rm T} p_{\rm c} = p_{\rm c}/p = 1$ in equation (2.7) was subtracted from the calculated $\kappa p_{\rm c}$ versus $T/T_{\rm c}$ data before making the plot in figure 10.8(*d*).

The $C_{\rm p}(T)$ predicted by equations (5.5) and (5.34*a*) is plotted for $p/p_{\rm c} = 1.0$–1.5 in figure 10.9(*a*). One sees that as p decreases towards $p_{\rm c}$ from above, a peak occurs at a temperature somewhat above $T_{\rm c}$ that develops into a divergent behavior at $T = T_{\rm c}$ when $p = p_{\rm c}$. The critical part of the divergent behavior $\frac{C_{\rm P}}{Nk_{\rm B}} - \frac{3}{2}$ for $p = p_{\rm c}$ is plotted

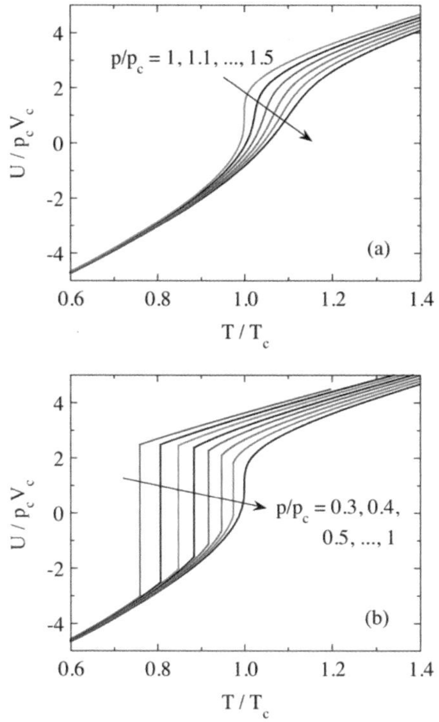

Figure 10.6. Equilibrium isobars of reduced internal energy $U/(p_c V_c)$ versus reduced temperature $\hat{\tau} = T/T_c$ calculated for (*a*) $\hat{p} = p/p_c \geqslant 1$ and (*b*) $\hat{p} \leqslant 1$, where the two-phase liquid–gas regions are indicated by vertical straight lines.

versus $|\frac{T}{T_c} - 1|$ for both $T/T_c < 1$ and $T/T_c > 1$ in a log–log plot in figure 10.9(*b*). The same critical behavior

$$\frac{C_p}{Nk_B} - \frac{3}{2} = 0.693(12)\left|\frac{T}{T_c} - 1\right|^{-0.667(1)}$$

is observed at the critical point for both $T \to T_c^+$ and $T \to T_c^-$, as shown, where the exponent and amplitude are equal to the analytical values of $-2/3$ and $1/3^{1/3}$ in equation (8.28) to within the error bars. From equation (5.33*b*), this critical exponent is consistent with the critical exponents of $-2/3$ determined above for both α and κ_T obtained on approaching the critical point at constant pressure versus temperature from either side of the critical point.

10.2.2 Pseudocritical curves and the Widom line in the supercritical region

a. Pseudocritical curves From figures 10.4–10.9, one sees that the molecular interactions have a large influence on n, ΔS, U, H, α, κ_T and C_p, even when the system is in the supercritical regime with $p > p_c$ and $T > T_c$. In particular, figure 10.4(*a*) shows

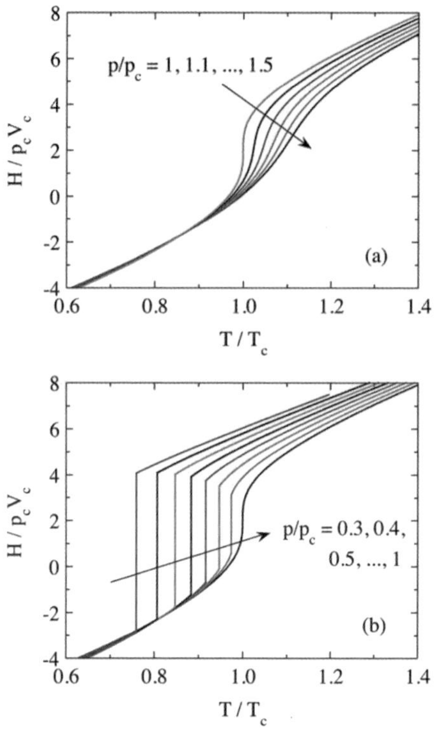

Figure 10.7. Equilibrium isobars of reduced enthalpy $H/(p_c V_c)$ versus reduced temperature $\hat{\tau} = T/T_c$ for (*a*) $\hat{p} = p/p_c \geqslant 1$ and (*b*) $\hat{p} \leqslant 1$, where the two-phase liquid–gas regions are indicated by vertical straight lines.

that the inflection point in each $\hat{n}$ versus $\hat{\tau}$ isobar separates a region of high (liquid-like) density from a region of low (gas-like) density, with the distinction being more dramatic as the critical point with $\hat{p} = 1$ is approached. The locus of these inflection points in the p–T plane is termed a pseudocritical curve (or ridge) that can be considered to define the crossover from the high- to low-density parts of the supercritical region. Other pseudocritical curves in the p–T plane can be defined by the locus of the maxima in α, κ_T and C_p isobars versus temperature and pressure. In this section we consider the pseudocritical curves associated with these four thermodynamic processes in the p–T plane.

From the equation of state (5.9), one can solve for $\hat{\tau}$ as a function of the constant isobar pressure $\hat{p}$ and $\hat{n}$ as

$$\hat{\tau} = \frac{3 - \hat{n}}{8\hat{n}}\left(\hat{p} + 3\hat{n}^2\right). \tag{10.7}$$

Substituting this expression for $\hat{\tau}$ into those for κ_T, α and C_p in equations (5.28*b*), (5.31*b*) and (5.34*b*), respectively, changes variables from $(\hat{\tau}, \hat{n})$ to $(\hat{p}, \hat{n})$, yielding

$$\frac{\alpha\tau_c}{k_B} = \frac{8\hat{n}/3}{\hat{p} - 3\hat{n}^2 + 2\hat{n}^3}, \tag{10.8a}$$

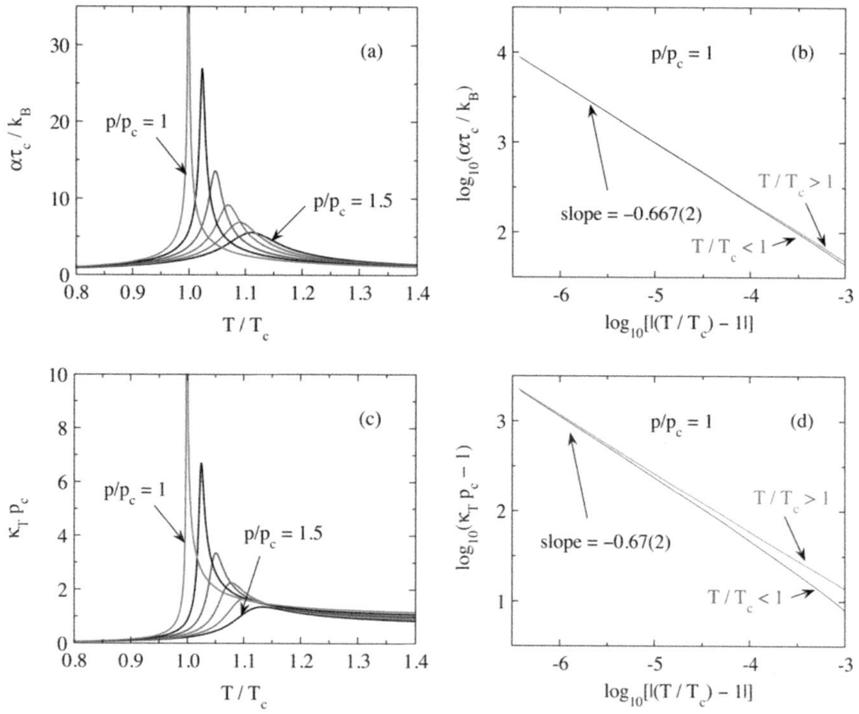

Figure 10.8. (*a*) Reduced volume thermal expansion coefficient $\alpha\tau_c/k_B$ versus reduced temperature $\hat{\tau} = T/T_c$ for reduced pressures $\hat{p} = p/p_c = 1$ to 1.5 in 0.1 increments obtained from equation (5.31*a*). The divergence of α for $p/p_c = 1$ is illustrated in (*b*) which shows a log–log plot of $\alpha\tau_c/k_B$ versus $|\hat{\tau} - 1|$ for both $\hat{\tau} \to 1^-$ (red) and $\hat{\tau} \to 1^+$ (blue). Both data sets show the same asymptotic behavior $\alpha\tau_c/k_B \propto |\hat{\tau} - 1|^{-2/3}$ for $\hat{\tau} \to 1$. Panels (*c*) and (*d*) show the same plots for the reduced isothermal compressibility $\kappa_T p_c$ versus $\hat{\tau}$ obtained using equation (5.28*a*) at the same pressures. The critical exponent in (*d*) is seen to be the same as for the thermal expansion coefficient in (*b*) to within the respective error bars.

$$\kappa_T p_c = \frac{(3 - \hat{n})/3}{\hat{p} - 3\hat{n}^2 + 2\hat{n}^3}, \tag{10.8b}$$

$$\frac{C_p - C_V}{Nk_B} = \frac{\hat{p} + 3\hat{n}^2}{\hat{p} - 3\hat{n}^2 + 2\hat{n}^3}. \tag{10.8c}$$

Setting the partial derivatives of these expressions with respect to $\hat{n}$ equal to zero gives the ridge equations

$$\hat{p}_{\alpha\,\mathrm{ridge}} = 4\hat{n}^3 - 3\hat{n}^2, \tag{10.9a}$$

$$\hat{p}_{\kappa_T\,\mathrm{ridge}} = \hat{n}[18 + \hat{n}(4\hat{n} - 21)], \tag{10.9b}$$

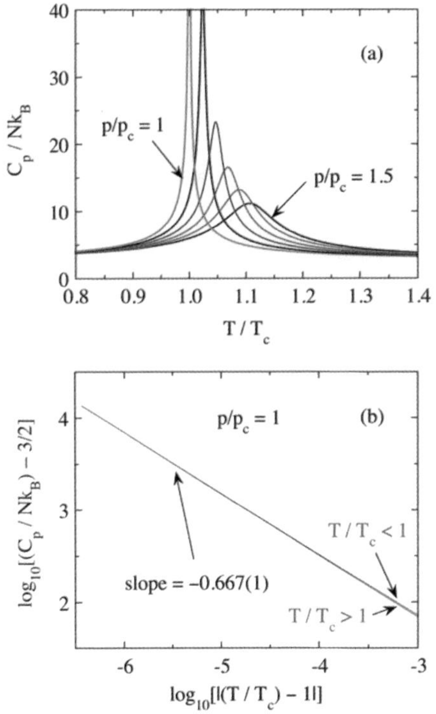

Figure 10.9. (*a*) Reduced heat capacity at constant pressure C_p/Nk_B versus reduced temperature $\hat{\tau} = T/T_c$ for reduced pressures $\hat{p} = p/p_c = 1$ to 1.5 in 0.1 increments. The divergence of C_p for $\hat{p} = 1$ is illustrated in (*b*) which shows a log–log plot of the critical part $\frac{C_p}{Nk_B} - \frac{3}{2}$ versus $|\hat{\tau} - 1|$ for both $\hat{\tau} \to 1^-$ (red) and $\hat{\tau} \to 1^+$ (blue). Both data sets show the same asymptotic behavior $\frac{C_p}{Nk_B} - \frac{3}{2} \propto |\hat{\tau} - 1|^{-2/3}$ for $\hat{\tau} \to 1$.

$$\hat{p}_{C_p\,\text{ridge}} = \frac{\hat{n}^3}{2 - \hat{n}}. \tag{10.9c}$$

These equations derived from isobars are different from equations (10.5) derived from isotherms.

From figure 10.4, the equation for the ridge associated with the number density $\hat{n}$ is obtained by setting the second derivative of $\hat{\tau}$ with respect to $\hat{n}$ equal to zero using equation (10.7), which gives the ridge equation for the number density inflection point as

$$\hat{p}_{n\,\text{ridge}} = \hat{n}^3. \tag{10.9d}$$

We wish to eventually plot $\hat{p}_{\text{ridge}}$ versus $\hat{\tau}$ in the supercritical region to compare with the coexistence curve in the subcritical region in the $\hat{p}$–$\hat{\tau}$ plane in figure 6.7. To do that one could calculate $\hat{\tau}(\hat{p}, \hat{n})$ and $\hat{p}_{\text{ridge}}(\hat{n})$ and then obtain $\hat{p}_{\text{ridge}}$ versus $\hat{\tau}$ with $\hat{n}$ as an implicit parameter using equations (10.7) and (10.9). A better procedure is to directly calculate $\hat{n}_{\text{ridge}}$ versus $\hat{p}$ from equations (10.9), which can be done exactly, and then substitute the result into equation (10.7) to obtain $\hat{\tau}_{\rm ridge}$ versus $\hat{p}$. Inverting equations (10.9) yields

$$\hat{n}_{\alpha\ {\rm ridge}} = \frac{1}{4}\left[1 + \frac{1}{\left(1 + 8\hat{p} - 4\sqrt{\hat{p} + 4\hat{p}^2}\right)^{1/3}} + \left(1 + 8\hat{p} - 4\sqrt{\hat{p} + 4\hat{p}^2}\right)^{1/3}\right], \tag{10.10a}$$

$$\hat{n}_{\kappa_{\rm T}\ {\rm ridge}} = \frac{1}{4}\left\{7 + 5\cos\left[\frac{1}{3}\arg\left(4\sqrt{(27 + \hat{p})(-17 + 4\hat{p})} - 91 - 8\hat{p}\right)\right]\right.$$
$$\left. - 5\sqrt{3}\sin\left[\frac{1}{3}\arg\left(4\sqrt{(27 + \hat{p})(-17 + 4\hat{p})} - 91 - 8\hat{p}\right)\right]\right\}, \tag{10.10b}$$

$$\hat{n}_{C_{\rm p}\ {\rm ridge}} = \frac{\left[9\hat{p} + \sqrt{3\left(27\hat{p}^2 + \hat{p}^3\right)}\right]^{1/3}}{3^{2/3}} - \frac{\hat{p}}{3^{1/3}\left[9\hat{p} + \sqrt{3\left(27\hat{p}^2 + \hat{p}^3\right)}\right]^{1/3}}, \tag{10.10c}$$

$$\hat{n}_{n\ {\rm ridge}} = \hat{p}^{1/3}. \tag{10.10d}$$

In equation (10.10*b*), arg(z) is the argument ϕ of the complex number $z = |z|{\rm e}^{{\rm i}\phi}$ and is a built-in `Mathematica` function.

To obtain $\hat{\tau}_{\rm ridge}$ versus $\hat{p}$ for the isobars of the above four thermodynamic functions, one inserts equations (10.10) into (10.7), respectively. The resulting ridge equations for α, $\kappa_{\rm T}$ and $C_{\rm p}$ are cumbersome and therefore not presented. However, in the case of the density ridge, one obtains the simple expression

$$\hat{\tau}_{n{\rm ridge}} = \frac{1}{8}\left(9\hat{p}^{1/3} - \hat{p}\right). \tag{10.11}$$

In order to compare these results for $\hat{\tau} \geqslant 1$ with the coexistence curve for $\hat{\tau} \leqslant 1$ in figure 6.7, we invert the axes of the ridges and plot the pressure of the ridges versus temperature in figure 10.10. These data agree with the respective ridge positions obtained from the plotted isobars versus temperature in figures 10.4(*a*), 10.8(*a*), 10.8(*c*) and 10.9(*a*). One sees from figure 10.10 that the ridges have a common behavior for $\hat{\tau} \to 1$ but diverge from each other at higher temperatures [3, 4]. Similar behaviors are observed for the Lennard-Jones fluid [5] and for supercritical water [6].

b. Widom line Xu *et al* [7] discussed the supercritical ridges for thermodynamic properties and stated that they asymptote to a common behavior for $\hat{\tau} \to 1$ because these ridges are expressible in terms of the same correlation length in that limit. They denoted the asymptotic line as the 'Widom line'. Following up on this observation, we obtained Taylor series expansions of $\hat{\tau}_{\rm ridge}$ about $\hat{p} - 1$ for α, $C_{\rm p}$ and n as

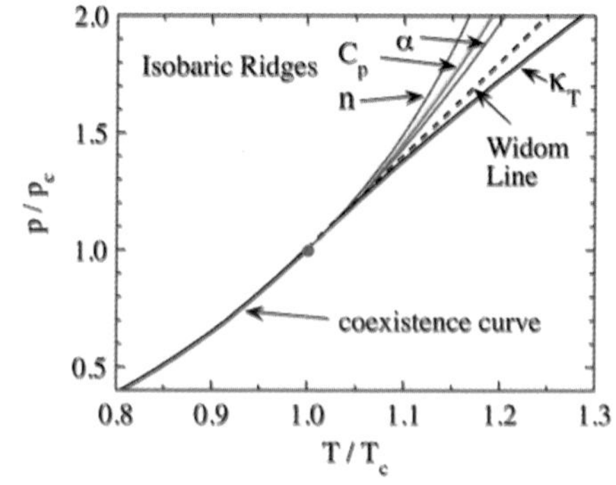

Figure 10.10. Reduced pressure $\hat{p} = p/p_c$ versus reduced temperature $\hat{\tau} = T/T_c$ for isobaric ridges of the thermal expansion coefficient α, heat capacity at constant pressure C_p, number density n and isothermal compressibility κ_T for the vdW fluid. The Widom line equation (10.13), to which these thermodynamic ridges asymptote as the critical point (filled red circle) is approached from above, is shown as the straight dashed black line. Also shown is a portion of the coexistence curve from figure 6.7. The Widom line is a linear extrapolation of the coexistence curve at $\hat{\tau} \leqslant 1$ to $\hat{\tau} \geqslant 1$ [3].

obtained from equations (10.7) and (10.10) or (10.11), respectively. To second order in $\hat{p} - 1$ they are

$$\hat{\tau}_{\alpha\ \text{ridge}} = 1 + \frac{1}{4}(\hat{p} - 1) - \frac{1}{16}(\hat{p} - 1)^2, \tag{10.12a}$$

$$\hat{\tau}_{C_p\ \text{ridge}} = 1 + \frac{1}{4}(\hat{p} - 1) - \frac{3}{32}(\hat{p} - 1)^2, \tag{10.12b}$$

$$\hat{\tau}_{n\ \text{ridge}} = 1 + \frac{1}{4}(\hat{p} - 1) - \frac{1}{8}(\hat{p} - 1)^2. \tag{10.12c}$$

The Widom line is the line to which these thermodynamic functions asymptote for $\hat{\tau} \to 1^+$ and $\hat{p} \to 1^+$, which for the vdW fluid is the linear temperature versus pressure relationship obtained by retaining only the first-order terms in equations (10.12). In agreement with the above statement by Xu *et al*, the linear term is indeed the same for each of the three ridges. From equations (10.12), the Widom line is therefore given for the vdW fluid in the $\hat{p}$ versus $\hat{\tau}$ projection as [3]

$$\hat{p}_{\text{Widom}} = 1 + 4(\hat{\tau} - 1) \qquad \text{(vdW fluid)}. \tag{10.13}$$

Comparing equation (10.13) with equation (7.7*c*) shows that the Widom line for the vdW fluid is the linear extrapolation to $\hat{\tau} \geqslant 1$ of the coexistence curve $p_X\,(\hat{\tau} \leqslant 1)$ in figure 6.7 [3]. The Widom line given by equation (10.13) is included as the dashed black straight line for $\hat{\tau} \geqslant 1$ in figure 10.10.

10.2.3 Results for pressures less than the critical pressure

Equilibrium $\Delta S/Nk_B$ versus $\hat{\tau}$ isobars calculated using equations (5.5) and (5.26) for pressures $\hat{p} \leqslant 1$, augmented by the above calculations of the gas–liquid coexistence region, are shown in figure 10.5(*b*). For $\hat{p} < 1$ a discontinuity in the entropy occurs at a temperature-dependent transition temperature T_X that decreases with decreasing $\hat{p}$ according to the pressure versus temperature phase diagram in figure 6.7. The change in entropy at the transition $\Delta S_X/Nk_B$ versus the reduced temperature T/T_c is plotted above in figure 7.6. Similar behaviors are found for the internal energy and enthalpy as shown in figures 10.6(*b*) and 10.7(*b*), respectively.

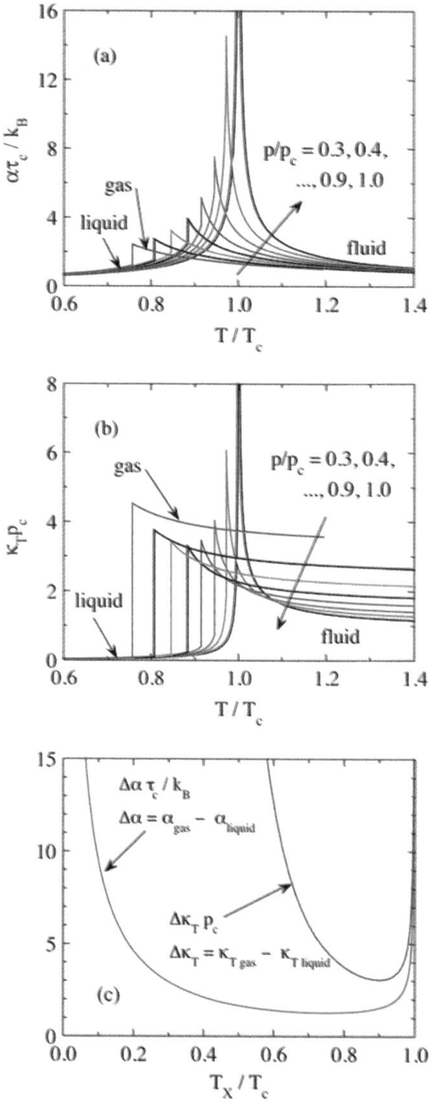

Figure 10.11. Isobars of the reduced (*a*) thermal expansion coefficient $\alpha\tau_c/k_B$ and (*b*) isothermal compressibility $\kappa_T p_c$ versus reduced temperature $\hat{\tau} = T/T_c \leqslant 1$. (*c*) The vertical discontinuities in $\alpha\tau_c/k_B$ and $\kappa_T p_c$ versus reduced transition temperature $\hat{\tau}_X = T_X/T_c$ predicted by equations (B.2*c*) and (B.1*c*) in appendix B, respectively.

The reduced thermal expansion coefficient $\alpha\tau_c/k_B$ and reduced isothermal compressibility $\kappa_T p_c$ versus reduced temperature $\hat{\tau} = T/T_c$ for several values of reduced pressure $\hat{p} = p/p_c$ are plotted in figures 10.11(*a*) and 10.11(*b*), respectively. Both quantities show discontinuous increases (jumps) at the first-order transition temperature $\hat{\tau}_X < 1$ from liquid to gas phases with increasing temperature. These data are for the pure gas and liquid phases on either side of the coexistence curve in figure 6.7.

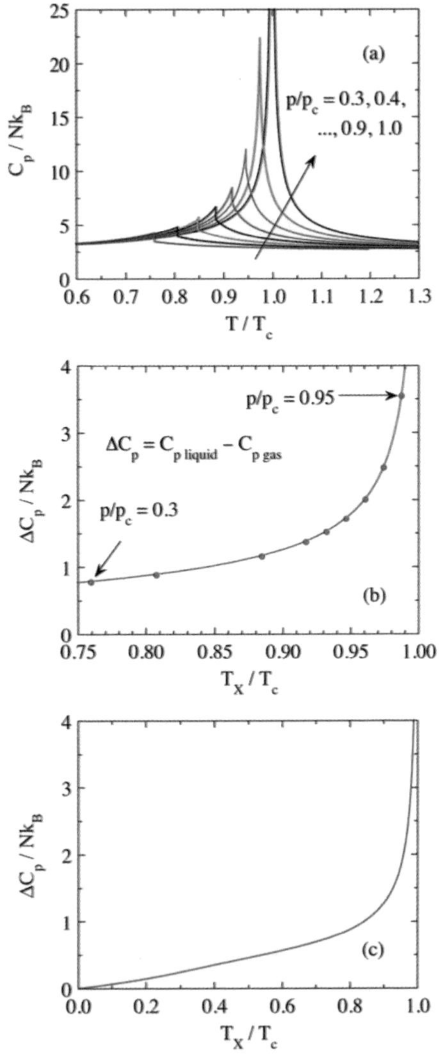

Figure 10.12. (*a*) Isobars of the reduced heat capacity at constant pressure $C_p/(Nk_B)$ versus reduced temperature $\hat{\tau} = T/T_c \leqslant 1$. (*b*) Heat capacity jump $\Delta C_p/(Nk_B)$ versus reduced transition temperature $\hat{\tau}_X = T_X/T_c$ obtained from isobars as in panel (*a*) (filled red circles). The solid curve is the prediction in equation (B.3*c*) in appendix B. (*c*) The prediction of equation (B.3*c*) for $\Delta C_p/Nk_B$ versus $\hat{\tau}_X$ over the full temperature range.

Remarkably, the jumps vary nonmonotonically with temperature for both quantitities. This is confirmed in figure 10.11(*c*) where the respective jumps calculated from the parametric solutions to them in equations (7.3*b*), (B.2*c*) and (B.1*c*) are plotted.

The reduced heat capacity at constant pressure $C_p/(Nk_B)$ versus reduced temperature $\hat{\tau}$ is shown in figure 10.12(*a*) for $\hat{p} = 0.3, 0.4, \ldots, 1$. The transition from pure gas to pure liquid on cooling below the reduced transition temperature $\hat{\tau}_X = T_X/T_c$ results in a peak in the heat capacity and a jump $\Delta C_p/Nk_B$ at $\hat{\tau}_X$. In addition, there is a latent heat at the transition that is not considered here. In contrast to the data for α and κ_T in figures 10.11(*a*) and 10.11(*b*), respectively, the positive jumps in C_p in figure 10.12 occur on cooling rather than on warming.

The heat capacity jumps obtained numerically from isobars as in figure 10.12(*a*) are plotted versus $\hat{\tau}_X$ as filled circles in figure 10.12(*b*), where it is seen to initially strongly decrease with decreasing $\hat{\tau}_X$ and then become much less dependent on $\hat{\tau}_X$. The exact parametric solution for $\Delta C_p/Nk_B$ versus $\hat{\tau}_X$ obtained from equations (7.3*b*) and (B.3*c*) is plotted as the solid curves in figures 10.12(*b*) and 10.12(*c*), where, in contrast to the jumps in κ_T and α in figure 10.11(*c*), $\Delta C_p/Nk_B$ decreases monotonically with decreasing $\hat{\tau}_X$ and goes to zero at $\hat{\tau}_X = 0$.

Representative values of the jumps versus $\hat{\tau}_X$ in κ_T, α and C_p on crossing the coexistence curve in figure 6.7, calculated from equations (B.1*c*), (B.2*c*) and (B.3*c*), respectively, are listed in table A.5 in appendix A.

References

[1] Nishikawa K, Kusano K, Arai A A and Morita T 2003 Density fluctuation of a van der Waals fluid in supercritical state *J. Chem. Phys.* **118** 1341

[2] Stanley H E 1971 *Introduction to Phase Transitions and Critical Phenomena* (New York: Oxford Science)

[3] Brazhkin V V and Ryzhov V N 2011 van der Waals supercritical fluid: exact formulas for special lines *J. Chem. Phys.* **135** 084503

[4] Luo J, Xu L, Lascaris E, Stanley H E and Buldyrev S V 2014 Behavior of the Widom line in critical phenomena *Phys. Rev. Lett.* **112** 135701

[5] May H-O and Mausbach P 2012 Riemannian geometry study of vapor–liquid phase equilibria and supercritical behavior of the Lennard-Jones fluid *Phys. Rev.* E **85** 031201

[6] Imre A R, Deiters U K, Kraska T and Tiselj I 2012 The pseudocritical regions for supercritical water *Nucl. Eng. Des.* **252** 179

[7] Xu L, Kumar P, Buldyrev S V, Chen S-H, Poole P H, Sciortino F and Stanley H E 2005 Relation between the Widom line and the dynamic crossover in systems with a liquid–liquid phase transition *Proc. Natl Acad. Sci. USA* **102** 16558

Advances in Thermodynamics of the van der Waals Fluid

David C Johnston

Chapter 11

Adiabatic free expansion and Joule–Thomson expansion

11.1 Adiabatic free expansion

In an adiabatic free expansion of a gas from an initial volume V_1 to a final volume V_2, the heat absorbed by the fluid Q and the work done by the fluid W during the expansion are both zero, so the change in the internal energy U of the fluid obtained from the first law of thermodynamics is

$$\Delta U \equiv U_2 - U_1 = Q - W = 0. \quad (11.1)$$

From the expression for the reduced internal energy of the vdW fluid in equation (5.18), one has

$$\frac{U_2 - U_1}{p_c V_c} = 4(\hat{\tau}_2 - \hat{\tau}_1) - 3\left(\frac{1}{\hat{V}_2} - \frac{1}{\hat{V}_1}\right). \quad (11.2)$$

Setting this equal to zero gives

$$\hat{\tau}_2 - \hat{\tau}_1 = \frac{3}{4}\left(\frac{1}{\hat{V}_2} - \frac{1}{\hat{V}_1}\right). \quad (11.3)$$

By definition of an expansion one has $\hat{V}_2 > \hat{V}_1$, yielding

$$\hat{\tau}_2 < \hat{\tau}_1, \quad (11.4)$$

so the adiabatic free expansion of a vdW fluid cools it. This contasts with an ideal gas where $\hat{\tau}_2 = \hat{\tau}_1$ because according to equation (2.4), U does not depend on volume for an ideal gas.

The above considerations are valid if there is no gas to liquid phase transition. To clarify this issue, in figure 11.1 are plots of $\hat{\tau}$ versus $\hat{V}$ obtained using equation (5.18)

doi:10.1088/978-1-627-05532-1ch11

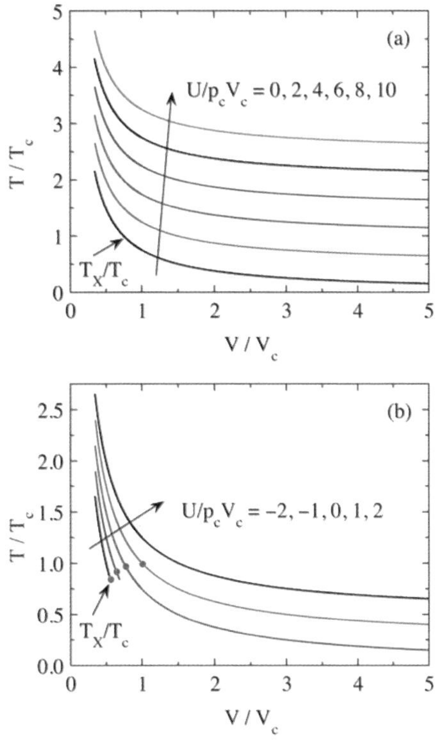

Figure 11.1. (*a*) Reduced temperature $\hat{\tau} = T/T_c$ versus reduced volume $\hat{V} = V/V_c$ at fixed values of reduced internal energy $U/(p_cV_c)$ for the ranges (*a*) $U/(p_cV_c) = 0$ to 10 and (*b*) −2 to 2. The filled red circles in (*b*) denote the phase transition from pure gas to pure liquid due to the cooling associated with the free expansion. In (*b*), the curves with $U/(p_cV_c) = -2$ and −1 terminate because the calculated pressure becomes negative at the ends of these plotted curves. For $U/(p_cV_c) = 0$ and 1, a wide range of possible initial and final volumes for the adiabatic expansion results in liquification of the expanding gas.

at the fixed values of $U/(p_cV_c)$ indicated in the figure. We have also calculated the pressure of the gas along each curve using equation (5.5) and compared it with the liquifaction pressure $\hat{p}_X(\hat{\tau})$ in figure 6.7. For a range of $U/(p_cV_c)$ between approximately 0 and 1, we find that the gas liquifies as it expands and cools, as shown by the filled red circles in figure 11.1(*b*). Thus under limited circumstances, adiabatic free expansion of the vdW gas can in principle liquify it.

However, we note the caveat discussed by Reif [1] that an adiabatic free expansion is an irreversible 'one-shot' expansion that necessarily has to cool the solid container that the gas is confined in, in order to reach thermal equilibrium. The container would likely have a substantial heat capacity compared to that of the gas. Therefore the actual amount of cooling of the gas is likely significantly smaller than calculated above. This limitation is eliminated in the steady-state expansion of a gas through a 'throttle' in a tube from high to low pressure as discussed in the next section, where in the steady state the walls of the tube on either side of the throttle have reached steady temperatures.

11.2 Joule–Thomson expansion

In Joule–Thomson (or Joule–Kelvin) expansion, gas at a high pressure p_1 passes through a constriction (throttle), which might be a porous plug or small valve, to a region with low pressure p_2 in a thermally insulated tube [1]. In such an expansion, the enthalpy H instead of the internal energy U is found to be constant during the expansion:

$$\frac{H_2(\hat{\tau}_2, \hat{p}_2)}{p_c V_c} = \frac{H_1(\hat{\tau}_1, \hat{p}_1)}{p_c V_c}. \tag{11.5}$$

Whether heating or cooling of the gas occurs due to the expansion depends on how $\hat{p}$ and $\hat{\tau}$ vary at constant H. Therefore it is useful to plot $\hat{\tau}$ versus $\hat{p}$ at fixed reduced H to characterize how the fluid temperature changes on passing from the high to the low pressure side of the throttle.

From equation (5.24) one can express the reduced temperature in terms of the reduced enthalpy and volume as

$$\hat{\tau}(H, \hat{V}) = \frac{3\hat{V} - 1}{4(5\hat{V} - 1)}\left(\frac{H}{p_c V_c} + \frac{6}{\hat{V}}\right). \tag{11.6}$$

However, one needs to plot $\hat{\tau}$ versus $\hat{p}$ instead of versus $\hat{V}$ at fixed H. Therefore in a parametric solution one calculates $\hat{\tau}$ versus $\hat{V}$ using equation (11.6) and $\hat{p}$ versus $\hat{V}$ using equation (5.5) and then obtains $\hat{\tau}$ versus $\hat{p}$ with $\hat{V}$ as an implicit parameter. We note that according to equation (5.25), a value of $H/(p_c V_c) = 2$ gives rise to a plot of $\hat{\tau}$ versus $\hat{p}$ that passes through the critical point $\hat{\tau} = \hat{p} = \hat{V} = 1$. Thus for $H/(p_c V_c) < 2$ one might expect that expansion of the vdW gas through a throttle could liquify the gas in a continuous steady-state process. This is confirmed below.

Shown in figure 11.2(*a*) are plots of reduced temperature $\hat{\tau}$ versus reduced pressure $\hat{p}$ at fixed values of reduced enthalpy $H/(p_c V_c)$ for $H/(p_c V_c) = -2$ to 10. Each curve has a smooth maximum. The point at the maximum of a curve is called an inversion point (labeled by a filled blue square) and the locus of these points versus H is known as the 'inversion curve'. If the high pressure p_1 is greater than the pressure of the inversion point, the gas would initially warm on expanding instead of cooling, whereas if p_1 is at a lower pressure than this, then the gas only cools as it expands through the throttle. Thus in using the Joule–Thomson expansion to cool a gas, one normally takes the high pressure p_1 to be at a lower pressure than the pressure of the inversion point. The low pressure p_2 can be adjusted according to the application.

The slope of T versus p at fixed H is [1]

$$\left(\frac{\partial T}{\partial p}\right)_H = \frac{V}{C_p}(T\alpha - 1). \tag{11.7}$$

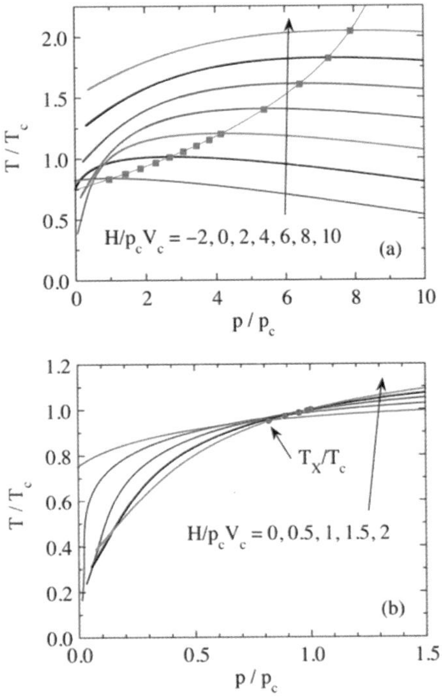

Figure 11.2. (*a*) Reduced temperature $\hat{\tau} = T/T_c$ versus reduced pressure $\hat{p} = p/p_c$ at fixed values of reduced enthalpy $H/(p_cV_c)$ for the range $H/(p_cV_c) = -2$ to 10. The filled blue squares are the inversion points with zero slope. The thin blue line through the blue squares is the analytic prediction for the inversion curve from equation (11.15*b*). (*b*) Explanded plots of $\hat{\tau}$ versus $\hat{p}$ for $H/(p_cV_c) = 0$ to 2. The filled red circle for each curve is $\hat{\tau}_X(\hat{p})$ for the given value of $H/(p_cV_c)$ that separates the region of pure gas to the right of the circle and pure liquid to the left, obtained from the coexistence curve $\hat{p}_X(\hat{\tau})$ in figure 6.7.

In terms of the reduced variables defined in equations (5.3), this becomes

$$\left(\frac{\partial \hat{\tau}}{\partial \hat{p}}\right)_H = \frac{3\hat{V}}{8\left[C_p/(Nk_B)\right]}\left[\hat{\tau}\left(\frac{\alpha\tau_c}{k_B}\right) - 1\right]. \tag{11.8}$$

Thus the inversion (I) point for a particular plot where the slope $\partial\hat{\tau}/\partial\hat{p}$ at fixed H changes from positive at low pressures to negative at high pressures is given by setting the right-hand side of equation (11.8) to zero, yielding

$$\hat{\tau}_I\left(H,\ \hat{V}\right) = \frac{1}{\alpha\tau_c/k_B} = \frac{3\left[4\hat{\tau}\left(H,\ \hat{V}\right)\hat{V}^3 - \left(3\hat{V} - 1\right)^2\right]}{4\left(3\hat{V} - 1\right)\hat{V}^2}, \tag{11.9}$$

where the second equality was obtained using the expression for α in equation (5.31*a*) and $\hat{\tau}(H,\ \hat{V})$ in the second equality is given in equation (11.6). Equation (11.9) allows an accurate determination of the inversion point by locating the value of $\hat{V}$ at which the calculated $\hat{\tau}(H, \hat{V})$ crosses $\hat{\tau}_{\rm I}(H, \hat{V})$ for the particular value of H.

One can also determine an analytic equation for the inversion curve of $\hat{\tau}_{\rm I}$ versus $\hat{p}$ and important points along it. By equating the temperatures $\hat{\tau}(H, \hat{V})$ in equation (11.6) and $\hat{\tau}_{\rm I}(H, \hat{V})$ in equation (11.9) one obtains the reduced volume $\hat{V}$ versus reduced enthalpy h at the inversion point as

$$\hat{V}_{\rm I} = \frac{15 + \sqrt{3(30 + h)}}{45 - h}, \tag{11.10}$$

where the reduced enthalpy is defined as

$$h \equiv \frac{H}{p_{\rm c} V_{\rm c}}. \tag{11.11}$$

Then $\hat{\tau}_{\rm I}$ is given in terms of h by inserting equation (11.10) into (11.6), yielding

$$\hat{\tau}_{\rm I} = \frac{1}{4}\left[42 + h - 4\sqrt{3(30 + h)}\right]. \tag{11.12}$$

The reduced inversion pressure versus h is obtained by inserting $\hat{V}_{\rm I}(h)$ in equation (11.10) and $\hat{\tau}_{\rm I}(h)$ in equation (11.12) into the equation of state (5.5), yielding

$$\hat{p} = 3\left[-75 - h + 8\sqrt{3(30 + h)}\right]. \tag{11.13}$$

Solving this expression for h gives the two solutions

$$h_{\pm} = 18 + \frac{9 - \hat{p}}{3} \pm 8(9 - \hat{p}). \tag{11.14}$$

Finally, inserting these two enthalpies into equation (11.12) and simplifying gives the two-branch solution for the inversion curve of $\hat{\tau}_{\rm I}(h)$ versus $\hat{p}$ as

$$\hat{\tau}_{\rm I} = 3 + \frac{9 - \hat{p}}{12} + \sqrt{9 - \hat{p}} \qquad (\hat{\tau}_{\rm I} \geqslant 3), \tag{11.15a}$$

$$\hat{\tau}_{\rm I} = 3 + \frac{9 - \hat{p}}{12} - \sqrt{9 - \hat{p}} \qquad (\hat{\tau}_{\rm I} \leqslant 3). \tag{11.15b}$$

The inverse relation was obtained in [2] as

$$\hat{p}_{\rm I} = 9\left(3 - 2\sqrt{\frac{\hat{\tau}}{3}}\right)\left(2\sqrt{\frac{\hat{\tau}}{3}} - 1\right), \tag{11.16}$$

which yields equations (11.15) on solving for $\hat{\tau}_{\rm I}(\hat{p})$.

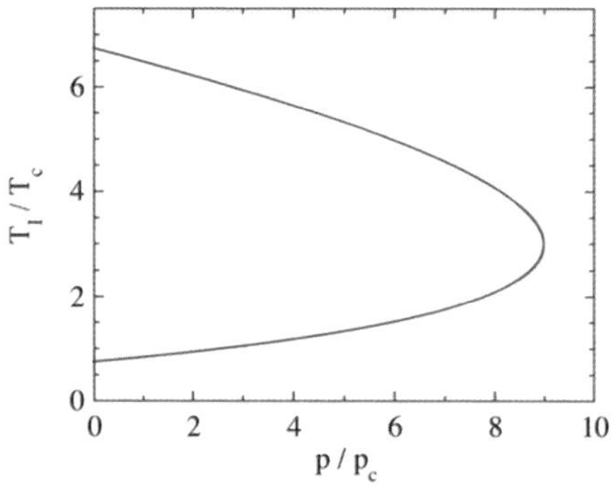

Figure 11.3. Reduced inversion temperature $\hat{\tau}_I = T_I/T_c$ versus reduced pressure $\hat{p} = p/p_c$ for Joule–Thomson expansion at constant enthalpy obtained using equations (11.15).

Important points along the inversion curve are [3]

$$\begin{aligned} \hat{p}_{max} &= 9, \\ \hat{\tau}_I(\hat{p} = 0) &= \frac{3}{4} \text{ and } \frac{27}{4}, \qquad h(\hat{p} = 0) = -3 \text{ and } 45, \\ \hat{\tau}_I(\hat{p} = \hat{p}_{max}) &= 3, \qquad h(\hat{p} = \hat{p}_{max}) = 18. \end{aligned} \tag{11.17}$$

These points are consistent with the inversion point data in figure 11.2(*a*). A plot of $\hat{\tau}_I$ versus $\hat{p}$ obtained using equation (11.15*b*) is shown in figure 11.2(*a*) and a plot using both of equations (11.15) is shown in figure 11.3. One notes from figure 11.3 that the curve is asymmetric with respect to a horizontal line through the apex of the curve.

Expanded plots of $\hat{\tau}$ versus $\hat{p}$ for $H/(p_c V_c) = 0$ to 2 are shown in figure 11.2(*b*) to emphasize the low-temperature and low-pressure region. On each curve is appended the data point $(\hat{\tau}, \hat{p}_X)$ (filled red circle) which is the corresponding point on the coexistence curve $\hat{p}_X(\hat{\tau})$ in figure 6.7. If the final pressure p_2 is to the left of the red circle for the curve, the fluid on the low-pressure side of the throttle is in the liquid phase, whereas if the final pressure is to the right of the red circle, the fluid is in the gas phase. Thus using Joule–Thomson expansion, one can continuously convert gas into liquid as the fluid cools within the throttle if one appropriately chooses the initial and final pressures.

References

[1] Reif F 1965 *Fundamentals of Statistical and Thermal Physics* (New York: McGraw-Hill)

[2] Le Vent S 2001 A summary of the properties of van der Waals fluids *Int. J. Mech. Eng. Educ.* **29** 257

[3] Kwok C K and Tilley D R 1979 A review of some thermodynamic properties of the van der Waals gas *Phys. Educ.* **14** 422

Appendix A

Tables of values

Table A.1. Phase coexistence points in the p–V–T phase space of the vdW fluid. The data columns versus temperature are labeled with the pressure and volume notations in figure 6.2. A symbol such as 'e – 01' means '$\times 10^{-1}$'.

T/T_c	p_X/p_c	V_G/V_c	V_F/V_c	V_D/V_c	V_C/V_c
1	1	1	1	1	1
0.99	9.605e−01	0.8309	0.8946	1.128	1.243e+00
0.98	9.219e−01	0.7755	0.8561	1.189	1.376e+00
0.97	8.843e−01	0.7376	0.8283	1.240	1.496e+00
0.96	8.476e−01	0.7082	0.8059	1.287	1.612e+00
0.95	8.119e−01	0.6841	0.7870	1.330	1.727e+00
0.94	7.771e−01	0.6637	0.7704	1.372	1.844e+00
0.93	7.432e−01	0.6459	0.7556	1.412	1.963e+00
0.92	7.102e−01	0.6302	0.7422	1.451	2.087e+00
0.91	6.782e−01	0.6161	0.7299	1.490	2.215e+00
0.90	6.470e−01	0.6034	0.7186	1.529	2.349e+00
0.89	6.167e−01	0.5918	0.7080	1.567	2.489e+00
0.88	5.874e−01	0.5811	0.6981	1.605	2.636e+00
0.87	5.589e−01	0.5712	0.6888	1.644	2.791e+00
0.86	5.312e−01	0.5620	0.6800	1.682	2.955e+00
0.85	5.045e−01	0.5534	0.6717	1.721	3.128e+00
0.84	4.786e−01	0.5453	0.6637	1.760	3.311e+00
0.83	4.535e−01	0.5377	0.6561	1.800	3.506e+00
0.82	4.293e−01	0.5306	0.6489	1.840	3.714e+00
0.81	4.059e−01	0.5238	0.6419	1.880	3.936e+00
0.80	3.834e−01	0.5174	0.6352	1.921	4.172e+00
0.79	3.616e−01	0.5113	0.6288	1.963	4.426e+00

(*Continued*)

doi:10.1088/978-1-627-05532-1ch12

Table A.1. *(Continued)*

T/T_c	p_X/p_c	V_G/V_c	V_F/V_c	V_D/V_c	V_C/V_c
0.78	3.406e−01	0.5055	0.6225	2.005	4.698e+00
0.77	3.205e−01	0.5000	0.6165	2.049	4.990e+00
0.76	3.011e−01	0.4947	0.6107	2.092	5.304e+00
0.75	2.825e−01	0.4896	0.6051	2.137	5.643e+00
0.74	2.646e−01	0.4848	0.5996	2.183	6.009e+00
0.73	2.475e−01	0.4801	0.5943	2.229	6.406e+00
0.72	2.311e−01	0.4756	0.5891	2.277	6.835e+00
0.71	2.154e−01	0.4713	0.5841	2.326	7.302e+00
0.70	2.005e−01	0.4672	0.5792	2.376	7.811e+00
0.69	1.862e−01	0.4632	0.5744	2.427	8.366e+00
0.68	1.726e−01	0.4593	0.5698	2.479	8.973e+00
0.67	1.597e−01	0.4556	0.5652	2.532	9.639e+00
0.66	1.475e−01	0.4520	0.5608	2.587	1.037e+01
0.65	1.358e−01	0.4485	0.5564	2.644	1.118e+01
0.64	1.249e−01	0.4451	0.5522	2.702	1.207e+01
0.63	1.145e−01	0.4419	0.5480	2.761	1.305e+01
0.62	1.047e−01	0.4387	0.5439	2.823	1.415e+01
0.61	9.550e−02	0.4356	0.5399	2.886	1.537e+01
0.60	8.687e−02	0.4326	0.5359	2.951	1.673e+01
0.59	7.878e−02	0.4297	0.5321	3.018	1.826e+01
0.58	7.123e−02	0.4269	0.5283	3.087	1.997e+01
0.57	6.419e−02	0.4241	0.5246	3.158	2.191e+01
0.56	5.764e−02	0.4214	0.5209	3.232	2.411e+01
0.55	5.158e−02	0.4188	0.5173	3.308	2.661e+01
0.54	4.598e−02	0.4163	0.5137	3.387	2.947e+01
0.53	4.081e−02	0.4138	0.5102	3.469	3.274e+01
0.52	3.607e−02	0.4114	0.5068	3.553	3.652e+01
0.51	3.174e−02	0.4091	0.5034	3.641	4.089e+01
0.50	2.779e−02	0.4068	0.5000	3.732	4.598e+01
0.49	2.420e−02	0.4045	0.4967	3.827	5.195e+01
0.48	2.097e−02	0.4023	0.4934	3.925	5.897e+01
0.47	1.806e−02	0.4002	0.4902	4.028	6.729e+01
0.46	1.545e−02	0.3981	0.4870	4.134	7.722e+01
0.45	1.313e−02	0.3960	0.4838	4.246	8.915e+01
0.44	1.108e−02	0.3940	0.4807	4.362	1.036e+02
0.43	9.283e−03	0.3921	0.4776	4.483	1.212e+02
0.42	7.710e−03	0.3901	0.4746	4.611	1.429e+02
0.41	6.347e−03	0.3883	0.4716	4.744	1.698e+02
0.40	5.175e−03	0.3864	0.4686	4.883	2.036e+02
0.39	4.175e−03	0.3846	0.4656	5.030	2.465e+02
0.38	3.332e−03	0.3828	0.4627	5.184	3.015e+02
0.37	2.627e−03	0.3811	0.4597	5.346	3.729e+02
0.36	2.044e−03	0.3794	0.4568	5.518	4.670e+02

Table A.1. (*Continued*)

T/T_c	p_X/p_c	V_G/V_c	V_F/V_c	V_D/V_c	V_C/V_c
0.35	1.567e−03	0.3777	0.4540	5.699	5.926e+02
0.34	1.183e−03	0.3761	0.4511	5.890	7.631e+02
0.33	8.785e−04	0.3745	0.4483	6.093	9.986e+02
0.32	6.401e−04	0.3729	0.4455	6.308	1.330e+03
0.31	4.569e−04	0.3713	0.4426	6.537	1.806e+03
0.30	3.188e−04	0.3698	0.4399	6.781	2.506e+03
0.29	2.170e−04	0.3683	0.4371	7.041	3.560e+03
0.28	1.437e−04	0.3668	0.4343	7.321	5.193e+03
0.27	9.225e−05	0.3654	0.4315	7.620	7.801e+03
0.26	5.723e−05	0.3639	0.4288	7.943	1.211e+04
0.25	3.417e−05	0.3625	0.4260	8.291	1.951e+04
0.24	1.953e−05	0.3612	0.4233	8.668	3.276e+04
0.23	1.063e−05	0.3598	0.4205	9.077	5.767e+04
0.22	5.477e−06	0.3585	0.4178	9.524	1.071e+05
0.21	2.647e−06	0.3571	0.4150	10.01	2.116e+05
0.20	1.189e−06	0.3558	0.4122	10.55	4.485e+05
0.19	4.909e−07	0.3546	0.4095	11.14	1.032e+06
0.18	1.836e−07	0.3533	0.4067	11.80	2.615e+06
0.17	6.113e−08	0.3521	0.4039	12.54	7.416e+06
0.16	1.773e−08	0.3508	0.4011	13.37	2.406e+07
0.15	4.360e−09	0.3496	0.3982	14.31	9.174e+07

Table A.2. Representative values for the quantities listed, calculated in terms of the implicit parameter y using Lekner's parametric solution [1]. The subscript X refers to a property associated with the coexistence curve in figure 6.7. $\hat{\tau}_X$ and $\hat{p}_X$ are the coordinates of the curve; $\Delta\hat{V}_X$ is the difference in volume of the coexisting gas (g) and liquid (l) phases; $\hat{n}_g$ and $\hat{n}_l$ are the number densities of the two coexisting phases, $\Delta\hat{n}_X$ is the difference in density between the coexisting phases, which is the order parameter for the gas–liquid transition; $\hat{n}_{ave}$ is the average of $\hat{n}_g$ and $\hat{n}_l$; $\Delta S_X/(Nk_B) = (S_g - S_l)/(Nk_B)$ is the entropy difference between the coexisting phases; and $L/(p_c V_c)$ is the latent heat (enthalpy) of vaporization. These data are complementary to those in table A.1.

y	$\hat{\tau}_X = T_X/T_c$	$\hat{p}_X = p_X/p_c$	$d\hat{p}_X/d\hat{\tau}_X$	$\Delta\hat{V}_X = (V_g - V_l)/V_c$	$\hat{n}_g = n_g/n_c$	$\hat{n}_l = n_l/n_c$	$\Delta\hat{n}_X = \hat{n}_l - \hat{n}_g$	$\hat{n}_{ave} = \frac{\hat{n}_g + \hat{n}_l}{2}$	$\frac{S_g - S_l}{Nk_B}$	$L/(p_c V_c)$
0	1	1	4	0	1	1	0	1	0	0
0.10000	0.99889	9.9557e−01	3.9893e+00	1.3369e−01	9.3384e−01	1.0671	0.13321	1.0004	0.20000	0.53274
0.11220	0.99860	9.9442e−01	3.9866e+00	1.5011e−01	9.2584e−01	1.0753	0.14944	1.0006	0.22440	0.59757
0.12589	0.99824	9.9298e−01	3.9831e+00	1.6857e−01	9.1689e−01	1.0845	0.16762	1.0007	0.25179	0.67025
0.14125	0.99779	9.9118e−01	3.9788e+00	1.8934e−01	9.0688e−01	1.0949	0.18801	1.0009	0.28251	0.75169
0.15849	0.99722	9.8891e−01	3.9733e+00	2.1274e−01	8.9569e−01	1.1065	0.21085	1.0011	0.31698	0.84292
0.17783	0.99650	9.8606e−01	3.9664e+00	2.3911e−01	8.8318e−01	1.1196	0.23644	1.0014	0.35566	0.94510
0.19953	0.99560	9.8248e−01	3.9578e+00	2.6887e−01	8.6921e−01	1.1343	0.26510	1.0018	0.39905	1.0595
0.22387	0.99446	9.7801e−01	3.9469e+00	3.0251e−01	8.5363e−01	1.1508	0.29717	1.0022	0.44774	1.1874
0.25119	0.99304	9.7240e−01	3.9333e+00	3.4059e−01	8.3626e−01	1.1693	0.33305	1.0028	0.50238	1.3304
0.28184	0.99126	9.6540e−01	3.9163e+00	3.8382e−01	8.1693e−01	1.1901	0.37315	1.0035	0.56368	1.4900
0.31623	0.98902	9.5666e−01	3.8949e+00	4.3301e−01	7.9544e−01	1.2134	0.41793	1.0044	0.63246	1.6680
0.35481	0.98622	9.4579e−01	3.8683e+00	4.8920e−01	7.7161e−01	1.2395	0.46786	1.0055	0.70963	1.8663
0.39811	0.98272	9.3231e−01	3.8350e+00	5.5365e−01	7.4523e−01	1.2687	0.52346	1.0070	0.79621	2.0865
0.44668	0.97835	9.1564e−01	3.7935e+00	6.2800e−01	7.1611e−01	1.3013	0.58524	1.0087	0.89337	2.3307
0.50119	0.97291	8.9514e−01	3.7420e+00	7.1432e−01	6.8408e−01	1.3378	0.65372	1.0109	1.0024	2.6006
0.56234	0.96615	8.7007e−01	3.6784e+00	8.1535e−01	6.4900e−01	1.3784	0.72941	1.0137	1.1247	2.8976
0.63096	0.95779	8.3964e−01	3.6000e+00	9.3476e−01	6.1078e−01	1.4235	0.81273	1.0171	1.2619	3.2231
0.70795	0.94748	8.0304e−01	3.5040e+00	1.0776e+00	5.6940e−01	1.4734	0.90404	1.0214	1.4159	3.5774
0.79433	0.93485	7.5951e−01	3.3872e+00	1.2507e+00	5.2494e−01	1.5284	1.0035	1.0267	1.5887	3.9604
0.89125	0.91947	7.0849e−01	3.2465e+00	1.4642e+00	4.7765e−01	1.5887	1.1111	1.0332	1.7825	4.3705

Table A.2. *(Continued)*

y	$\hat{\tau}_X = T_X/T_c$	$\hat{p}_X = p_X/p_c$	$d\hat{p}_X/d\hat{\tau}_X$	$\Delta\hat{V}_X = (V_g - V_l)/V_c$	$\hat{n}_g = n_g/n_c$	$\hat{n}_l = n_l/n_c$	$\Delta\hat{n}_X = \hat{n}_l - \hat{n}_g$	$\hat{n}_{ave} = \frac{\hat{n}_g + \hat{n}_l}{2}$	$\frac{S_g - S_l}{Nk_B}$	$L/(p_c V_c)$
1.0000	0.90088	6.4971e−01	3.0787e+00	1.7324e+00	4.2793e−01	1.6543	1.2264	1.0411	2.0000	4.8047
1.1220	0.87864	5.8343e−01	2.8812e+00	2.0769e+00	3.7642e−01	1.7251	1.3487	1.0508	2.2440	5.2579
1.2589	0.85233	5.1066e−01	2.6528e+00	2.5310e+00	3.2401e−01	1.8008	1.4768	1.0624	2.5179	5.7228
1.4125	0.82165	4.3327e−01	2.3938e+00	3.1471e+00	2.7183e−01	1.8806	1.6088	1.0762	2.8251	6.1900
1.5849	0.78646	3.5408e−01	2.1075e+00	4.0108e+00	2.2124e−01	1.9638	1.7425	1.0925	3.1698	6.6477
1.7783	0.74684	2.7672e−01	1.8004e+00	5.2677e+00	1.7374e−01	2.0489	1.8751	1.1113	3.5566	7.0831
1.9953	0.70320	2.0516e−01	1.4831e+00	7.1752e+00	1.3083e−01	2.1345	2.0036	1.1326	3.9905	7.4830
2.2387	0.65627	1.4305e−01	1.1693e+00	1.0211e+01	9.3795e−02	2.2188	2.1250	1.1563	4.4774	7.8358
2.5119	0.60711	9.2953e−02	8.7509e−01	1.5309e+01	6.3517e−02	2.3003	2.2368	1.1819	5.0238	8.1333
2.8184	0.55698	5.5764e−02	6.1578e−01	2.4410e+01	4.0272e−02	2.3773	2.3370	1.2088	5.6368	8.3722
3.1623	0.50721	3.0600e−02	4.0325e−01	4.1824e+01	2.3679e−02	2.4485	2.4249	1.2361	6.3246	8.5544
3.5481	0.45904	1.5218e−02	2.4302e−01	7.7868e+01	1.2777e−02	2.5133	2.5006	1.2631	7.0963	8.6867
3.9811	0.41346	6.7958e−03	1.3318e−01	1.5943e+02	6.2571e−03	2.5714	2.5651	1.2888	7.9621	8.7788
4.4668	0.37112	2.6990e−03	6.5526e−02	3.6357e+02	2.7476e−03	2.6227	2.6200	1.3127	8.9337	8.8412
5.0119	0.33233	9.4325e−04	2.8554e−02	9.3610e+02	1.0678e−03	2.6678	2.6668	1.3344	10.024	8.8833
5.6234	0.29716	2.8653e−04	1.0860e−02	2.7617e+03	3.6204e−04	2.7073	2.7069	1.3538	11.247	8.9122
6.3096	0.26545	7.4573e−05	3.5467e−03	9.4880e+03	1.0539e−04	2.7418	2.7417	1.3710	12.619	8.9327
7.0795	0.23699	1.6350e−05	9.7698e−04	3.8647e+04	2.5875e−05	2.7720	2.7720	1.3860	14.159	8.9479
7.9433	0.21148	2.9615e−06	2.2247e−04	1.9043e+05	5.2513e−06	2.7985	2.7985	1.3992	15.887	8.9594
8.9125	0.18867	4.3334e−07	4.0941e−05	1.1610e+06	8.6131e−07	2.8217	2.8217	1.4108	17.825	8.9682
10.000	0.16828	4.9947e−08	5.9361e−06	8.9845e+06	1.1130e−07	2.8421	2.8421	1.4211	20.000	8.9751
11.220	0.15007	4.4071e−09	6.5899e−07	9.0807e+07	1.1012e−08	2.8601	2.8601	1.4300	22.440	8.9804
12.589	0.13381	2.8826e−10	5.4240e−08	1.2379e+09	8.0783e−10	2.8759	2.8759	1.4380	25.179	8.9846
14.125	0.11930	1.3482e−11	3.1924e−09	2.3598e+10	4.2376e−11	2.8899	2.8899	1.4450	28.251	8.9879
15.849	0.10636	4.3290e−13	1.2901e−10	6.5518e+11	1.5263e−12	2.9023	2.9023	1.4511	31.698	8.9904

Table A.3. Representative values of the normalized heat capacity at constant volume $C_V/(Nk_B)$ versus Lekner's implicit parameter y and reduced temperature $\hat{\tau} = T/T_c$ for thermodynamic paths along isochores with reduced volumes $\hat{V} \equiv V/V_c = 1/2$, 1 and 2 calculated using equations (7.3) and (7.21). Finite discontinuities in $C_V/(Nk_B)$ occur at the respective second-order mixed-phase to single-phase transition temperatures.

		$C_V/(Nk_B)$		
y	$\hat{\tau}_X$	$\hat{V} = 1/2$	$\hat{V} = 1$	$\hat{V} = 2$
	1.2	1.500	1.500	1.500
0	1	1.500	1.500	1.500
0	1		6.000	
0.2818	0.9913		5.956	
0.3981	0.9827		5.913	
0.5012	0.9729		5.863	
0.5623	0.9662		5.829	
0.6310	0.9578		5.786	
0.7079	0.9475		5.734	
0.7943	0.9349		5.669	
	0.9270			1.500
	0.9270			8.809
0.8913	0.9195		5.591	8.726
1.000	0.9009		5.495	8.522
1.122	0.8786		5.381	8.277
1.259	0.8523		5.244	7.985
1.413	0.8217		5.085	7.643
1.585	0.7865		4.901	7.248
	0.7700	1.500		
	0.7700	3.690		
1.778	0.7468	3.638	4.693	6.802
1.995	0.7032	3.540	4.463	6.310
2.239	0.6563	3.435	4.218	5.782
2.512	0.6071	3.326	3.963	5.238
2.818	0.5570	3.216	3.710	4.699
3.162	0.5072	3.110	3.471	4.194
3.548	0.4590	3.012	3.258	3.749
3.981	0.4135	2.926	3.079	3.385
4.467	0.3711	2.853	2.940	3.112
5.012	0.3323	2.794	2.837	2.923
5.623	0.2972	2.747	2.766	2.803
6.310	0.2655	2.710	2.717	2.731
7.079	0.2370	2.680	2.682	2.686
7.943	0.2115	2.655	2.656	2.657

Table A.3. (*Continued*)

y	$\hat{\tau}_X$	$C_V/(Nk_B)$		
		$\hat{V} = 1/2$	$\hat{V} = 1$	$\hat{V} = 2$
8.913	0.1887	2.635	2.635	2.635
10.00	0.1683	2.618	2.618	2.618
11.22	0.1501	2.603	2.603	2.603
14.13	0.1193	2.579	2.579	2.579
17.78	0.09482	2.561	2.561	2.561
19.95	0.08452	2.554	2.554	2.554
22.39	0.07534	2.548	2.548	2.548
28.18	0.05986	2.537	2.537	2.537
50.12	0.03367	2.521	2.521	2.521
∞	0	5/2	5/2	5/2

Table A.4. Representative temperatures and volumes versus pressure associated with superheating and supercooling of the vdW fluid. The data columns are labeled with the notations in figure 9.1.

p/p_c	T_{SC}/T_c	T_X/T_c	T_{SH}/T_c	V_7/V_c	V_3/V_c	V_5/V_c	V_8/V_c	V_4/V_c	V_6/V_c	V_2/V_c
0.0002	0.0183	0.2880	0.8438	0.3680	3.836e+03	3.542	0.3352	1.221e+02	0.6667	1.125e+04
0.0004	0.0259	0.3062	0.8438	0.3707	2.038e+03	3.308	0.3359	8.627e+01	0.6667	5.624e+03
0.0006	0.0317	0.3180	0.8438	0.3726	1.410e+03	3.172	0.3365	7.037e+01	0.6667	3.749e+03
0.0008	0.0365	0.3270	0.8439	0.3740	1.087e+03	3.075	0.3370	6.090e+01	0.6667	2.812e+03
0.0100	0.1274	0.4341	0.8450	0.3929	1.135e+02	2.243	0.3470	1.698e+01	0.6677	2.243e+02
0.0120	0.1393	0.4446	0.8453	0.3949	9.656e+01	2.185	0.3483	1.547e+01	0.6679	1.868e+02
0.0140	0.1502	0.4539	0.8455	0.3968	8.425e+01	2.136	0.3496	1.429e+01	0.6681	1.600e+02
0.0160	0.1603	0.4622	0.8458	0.3985	7.487e+01	2.095	0.3509	1.335e+01	0.6683	1.400e+02
0.0180	0.1697	0.4698	0.8460	0.4001	6.748e+01	2.058	0.3520	1.256e+01	0.6685	1.243e+02
0.0200	0.1786	0.4768	0.8463	0.4016	6.148e+01	2.025	0.3531	1.190e+01	0.6687	1.118e+02
0.0300	0.2174	0.5057	0.8475	0.4081	4.298e+01	1.901	0.3581	9.648e+00	0.6697	7.433e+01
0.0400	0.2496	0.5283	0.8488	0.4134	3.333e+01	1.814	0.3624	8.305e+00	0.6707	5.558e+01
0.0500	0.2777	0.5473	0.8500	0.4181	2.735e+01	1.749	0.3664	7.388e+00	0.6717	4.433e+01
0.0600	0.3028	0.5637	0.8513	0.4224	2.327e+01	1.696	0.3701	6.711e+00	0.6728	3.682e+01
0.0700	0.3257	0.5783	0.8526	0.4264	2.029e+01	1.652	0.3736	6.184e+00	0.6738	3.147e+01
0.0800	0.3468	0.5915	0.8539	0.4301	1.801e+01	1.614	0.3770	5.758e+00	0.6749	2.745e+01
0.0900	0.3665	0.6037	0.8552	0.4337	1.621e+01	1.581	0.3802	5.406e+00	0.6759	2.432e+01
0.1000	0.3849	0.6150	0.8564	0.4371	1.474e+01	1.552	0.3834	5.107e+00	0.6770	2.182e+01
0.1200	0.4188	0.6354	0.8590	0.4436	1.251e+01	1.502	0.3895	4.626e+00	0.6792	1.806e+01
0.1400	0.4496	0.6536	0.8616	0.4498	1.087e+01	1.461	0.3953	4.251e+00	0.6814	1.538e+01
0.1800	0.5039	0.6855	0.8669	0.4614	8.633e+00	1.395	0.4067	3.696e+00	0.6860	1.180e+01
0.2000	0.5283	0.6997	0.8695	0.4671	7.828e+00	1.368	0.4122	3.483e+00	0.6884	1.055e+01
0.2200	0.5512	0.7130	0.8722	0.4726	7.159e+00	1.343	0.4178	3.299e+00	0.6908	9.524e+00
0.2400	0.5728	0.7255	0.8749	0.4781	6.595e+00	1.322	0.4233	3.138e+00	0.6933	8.668e+00
0.2600	0.5933	0.7374	0.8776	0.4835	6.111e+00	1.302	0.4288	2.995e+00	0.6959	7.943e+00
0.2800	0.6128	0.7486	0.8803	0.4890	5.691e+00	1.283	0.4343	2.868e+00	0.6985	7.321e+00
0.3000	0.6314	0.7594	0.8831	0.4944	5.323e+00	1.267	0.4399	2.753e+00	0.7011	6.781e+00

Table A.4. (*Continued*)

p/p_c	T_{SC}/T_c	T_X/T_c	T_{SH}/T_c	V_7/V_c	V_3/V_c	V_5/V_c	V_8/V_c	V_4/V_c	V_6/V_c	V_2/V_c
0.3200	0.6491	0.7698	0.8859	0.4999	4.997e+00	1.251	0.4455	2.649e+00	0.7039	6.308e+00
0.3400	0.6661	0.7797	0.8886	0.5053	4.707e+00	1.237	0.4511	2.553e+00	0.7067	5.890e+00
0.3600	0.6824	0.7892	0.8915	0.5109	4.446e+00	1.223	0.4568	2.466e+00	0.7095	5.518e+00
0.3800	0.6981	0.7985	0.8943	0.5165	4.210e+00	1.210	0.4627	2.385e+00	0.7125	5.184e+00
0.4000	0.7132	0.8074	0.8971	0.5221	3.996e+00	1.198	0.4686	2.310e+00	0.7155	4.883e+00
0.4400	0.7418	0.8245	0.9029	0.5337	3.620e+00	1.176	0.4807	2.174e+00	0.7218	4.362e+00
0.4800	0.7685	0.8406	0.9088	0.5457	3.301e+00	1.157	0.4934	2.055e+00	0.7285	3.925e+00
0.5200	0.7936	0.8558	0.9148	0.5583	3.025e+00	1.139	0.5068	1.948e+00	0.7357	3.553e+00
0.5600	0.8170	0.8704	0.9209	0.5715	2.785e+00	1.122	0.5209	1.852e+00	0.7433	3.232e+00
0.6000	0.8391	0.8843	0.9271	0.5856	2.571e+00	1.107	0.5359	1.763e+00	0.7516	2.951e+00
0.6400	0.8600	0.8977	0.9334	0.6006	2.380e+00	1.093	0.5522	1.682e+00	0.7605	2.702e+00
0.6800	0.8796	0.9106	0.9399	0.6169	2.207e+00	1.080	0.5698	1.607e+00	0.7702	2.479e+00
0.7200	0.8982	0.9230	0.9465	0.6347	2.049e+00	1.068	0.5891	1.535e+00	0.7810	2.277e+00
0.7600	0.9157	0,9350	0.9533	0.6545	1.903e+00	1.056	0.6107	1.468e+00	0.7930	2.092e+00
0.8000	0.9322	0.9466	0.9603	0.6769	1.766e+00	1.046	0.6352	1.403e+00	0.8066	1.921e+00
0.8400	0.9478	0.9579	0.9676	0.7027	1.636e+00	1.035	0.6637	1.339e+00	0.8224	1.760e+00
0.8800	0.9624	0.9688	0.9750	0.7338	1.510e+00	1.026	0.6981	1.276e+00	0.8414	1.605e+00
0.9200	0.9761	0.9795	0.9828	0.7733	1.382e+00	1.017	0.7422	1.210e+00	0.8655	1.451e+00
0.9600	0.9887	0.9899	0.9910	0.8300	1.245e+00	1.008	0.8059	1.137e+00	0.8998	1.287e+00
1	1	1	1	1	1	1	1	1	1	1

Table A.5. Representative values of the jumps in three properties on crossing the coexistence curve in figure 6.7. The reduced temperatures $\hat{\tau}_X$ (and y values) are the same as in table A.2. The quantities listed are the jumps in the reduced isothermal compressibility κ_T, thermal expansion coefficient α and heat capacity at constant pressure C_p calculated from equations (B.1*c*), (B.2*c*) and (B.3*c*), as shown in figures 10.11 and 10.12.

$\hat{\tau}_X$	$\Delta\kappa_T p_c = (\kappa_{Tg} - \kappa_{Tl})p_c$	$\Delta\alpha\ \tau_c/k_B = \frac{(\alpha_g - \alpha_l)\tau_c}{k_B}$	$\Delta C_p/(Nk_B) = \frac{C_{pl} - C_{pg}}{Nk_B}$
0.99889	1.8110e+01	12.035	1.2007e+01
0.99860	1.6164e+01	10.734	1.0703e+01
0.99824	1.4418e+01	9.5642	9.5408e+00
0.99779	1.2882e+01	8.5343	8.5053e+00
0.99722	1.1513e+01	7.6148	7.5827e+00
0.99650	1.0297e+01	6.7967	6.7607e+00
0.99560	9.2177e+00	6.0688	6.0283e+00
0.99446	8.2611e+00	5.4214	5.3760e+00
0.99304	7.4144e+00	4.8460	4.7950e+00
0.99126	6.6665e+00	4.3350	4.2776e+00
0.98902	6.0077e+00	3.8815	3.8169e+00
0.98622	5.4293e+00	3.4795	3.4069e+00
0.98272	4.9242e+00	3.1238	3.0420e+00
0.97835	4.4860e+00	2.8096	2.7175e+00
0.97291	4.1097e+00	2.5329	2.4290e+00
0.96615	3.7913e+00	2.2899	2.1726e+00
0.95779	3.5282e+00	2.0776	1.9449e+00
0.94748	3.3190e+00	1.8933	1.7428e+00
0.93485	3.1645e+00	1.7347	1.5636e+00
0.91947	3.0676e+00	1.5998	1.4048e+00
0.90088	3.0351e+00	1.4872	1.2642e+00
0.87864	3.0789e+00	1.3958	1.1398e+00
0.85233	3.2197e+00	1.3250	1.0295e+00
0.82165	3.4926e+00	1.2747	9.3181e−01
0.78646	3.9586e+00	1.2451	8.4488e−01
0.74684	4.7256e+00	1.2374	7.6714e−01
0.70320	5.9936e+00	1.2531	6.9701e−01
0.65627	8.1520e+00	1.2947	6.3297e−01
0.60711	1.2003e+01	1.3654	5.7362e−01
0.55698	1.9313e+01	1.4688	5.1776e−01
0.50721	3.4261e+01	1.6090	4.6451e−01
0.45904	6.7577e+01	1.7899	4.1347e−01
0.41346	1.4940e+02	2.0145	3.6480e−01
0.37112	3.7328e+02	2.2849	3.1908e−01
0.33233	1.0636e+03	2.6019	2.7714e−01
0.29716	3.4944e+03	2.9662	2.3968e−01
0.26545	1.3415e+04	3.3788	2.0700e−01

Table A.5. (*Continued*)

$\hat{\tau}_X$	$\Delta\kappa_T p_c = (\kappa_{Tg} - \kappa_{Tl})p_c$	$\Delta\alpha\ \tau_c/k_B = \dfrac{(\alpha_g - \alpha_l)\tau_c}{k_B}$	$\Delta C_p/(Nk_B) = \dfrac{C_{pl} - C_{pg}}{Nk_B}$
0.23699	6.1168e+04	3.8425	1.7898e−01
0.21148	3.3768e+05	4.3618	1.5514e−01
0.18867	2.3077e+06	4.9427	1.3490e−01
0.16828	2.0021e+07	5.5928	1.1765e−01
0.15007	2.2691e+08	6.3207	1.0288e−01
0.13381	3.4691e+09	7.1361	9.0177e−02
0.11930	7.4175e+10	8.0499	7.9206e−02
0.10636	2.3100e+12	9.0743	6.9692e−02

Reference

[1] Lekner J 1982 Parametric solution of the van der Waals liquid–vapor coexistence curve *Am. J. Phys.* **50** 161

Appendix B

Formulas for the discontinuities in isothermal compressibility, thermal expansion and heat capacity versus temperature at constant pressure on crossing the liquid–gas coexistence curve

The following equations for the discontinuities in the isothermal compressibility κ_T, volume thermal expansion coefficient α and heat capacity at constant pressure C_p versus reduced temperature $\hat{\tau}_X$ on crossing the coexistence curve in figure 6.7 are obtained from the calculation of these properties in terms of the parameter y together with $\hat{\tau}_X(y)$ in equation (7.3*b*) with y being an implicit parameter.

$$\kappa_{Tg}p_c = \frac{4e^{6y}\sinh^2(y)\left[2y^2 - 1 + \cosh(2y) - 2y\sinh(2y)\right]^2}{27\left[1 + e^{2y}(y-1) + y\right]^2\left[e^{6y} + e^{2y}(7-4y) - 2y - 3 + e^{4y}\left(-5 + 6y - 8y^2\right)\right]}, \tag{B.1a}$$

$$\kappa_{Tl}p_c = \frac{-e^{-y}\sinh^2(y)\left[1 - 2y^2 - \cosh(2y) + 2y\sinh(2y)\right]^2}{\left[\begin{array}{l}54\left[y\cosh(y) - \sinh(y)\right]^2\left\{\left(4y^2 - 1 + y\right)\cosh(y) - (y-1)\cosh(3y)\right. \\ \qquad \left. - 2y[3 + 2y + \cosh(2y)]\sinh(y) + 8\sinh^3(y)\right\}\end{array}\right]}, \tag{B.1b}$$

doi:10.1088/978-1-627-05532-1ch13

$$\begin{aligned}
\Delta\kappa_{\mathrm{T}} p_{\mathrm{c}} &\equiv \left(\kappa_{\mathrm{T}_g} - \kappa_{\mathrm{T}_l}\right) p_{\mathrm{c}} = \frac{\Delta\kappa_{\mathrm{T}}^{(1)}}{\Delta\kappa_{\mathrm{T}}^{(2)}}, \\
\Delta\kappa_{\mathrm{T}}^{(1)} &= 4\sinh^4(y)\left\{2y[2+\cosh(2y)] - 3\sinh(2y)\right\} \\
&\quad \times \left[2y^2 - 1 + \cosh(2y) - 2y\sinh(2y)\right]^2, \\
\Delta\kappa_{\mathrm{T}}^{(2)} &= 27\left[y\cosh(y) - \sinh(y)\right]^2 \left\{42 + 12y^2 + 32y^4 - \left(61 + 48y^2\right)\cosh(2y)\right. \\
&\quad + \left(22 + 36y^2\right)\cosh(4y) - 3\cosh(6y) - 128y^3\cosh^3(y)\sinh(y) \\
&\quad \left. - 16y[9\cosh(y) - \cosh(3y)]\sinh^3(y)\right\}.
\end{aligned} \tag{B.1c}$$

$$\begin{aligned}
\frac{\alpha_g \tau_{\mathrm{c}}}{k_{\mathrm{B}}} &= \frac{\alpha_g^{(1)}}{\alpha_g^{(2)}}, \\
\alpha_g^{(1)} &= 8\sinh^2(y)\left[2y^2 - 1 + \cosh(2y) - 2y\sinh(2y)\right]^2 \\
\alpha_g^{(2)} &= 27[y\cosh(y) - \sinh(y)][2y - \sinh(2y)] \\
&\quad \times \left\{[y(4y-1) - 1]\cosh(y) + (1+y)\cosh(3y)\right. \\
&\quad \left. + [6 + y(4y-5)]\sinh(y) - (2+y)\sinh(3y)\right\},
\end{aligned} \tag{B.2a}$$

$$\begin{aligned}
\frac{\alpha_l \tau_{\mathrm{c}}}{k_{\mathrm{B}}} &= \frac{\alpha_l^{(1)}}{\alpha_l^{(2)}}, \\
\alpha_l^{(1)} &= 8\sinh^2(y)\left[2y^2 - 1 + \cosh(2y) - 2y\sinh(2y)\right]^2, \\
\alpha_l^{(2)} &= 27[y\cosh(y) - \sinh(y)] \\
&\quad \times \left\{\left(y - 1 + 4y^2\right)\cosh(y) - (y-1)\cosh(3y)\right. \\
&\quad \left. - 2y[3 + 2y + \cosh(2y)]\sinh(y) + 8\sinh^3(y)\right\}[2y - \sinh(2y)],
\end{aligned} \tag{B.2b}$$

$$\begin{aligned}
\frac{\Delta\alpha_l \tau_{\mathrm{c}}}{k_{\mathrm{B}}} &\equiv \frac{(\alpha_g - \alpha_l)\tau_{\mathrm{c}}}{k_{\mathrm{B}}} = \frac{\Delta\alpha^{(1)}}{\Delta\alpha^{(2)}}, \\
\Delta\alpha^{(1)} &= 64\sinh^3(y)\left[1 - 2y^2 - \cosh(2y) + 2y\sinh(2y)\right]^2 \\
&\quad \times \left[2\cosh(2y) - y\sinh(2y) - 2 - 2y^2\right], \\
\Delta\alpha^{(2)} &= 27[y\cosh(y) - \sinh(y)]\left\{42 + 12y^2 + 32y^4 - \left(61 + 48y^2\right)\cosh(2y)\right. \\
&\quad + \left(22 + 36y^2\right)\cosh(4y) - 3\cosh(6y) - 128y^3\cosh^3(y)\sinh(y) \\
&\quad \left. + 16y[\cosh(3y) - 9\cosh(y)]\sinh^3(y)\right\}[2y - \sinh(2y)].
\end{aligned} \tag{B.2c}$$

$$\frac{C_{\mathrm{pg}}}{Nk_{\mathrm{B}}} = \frac{3}{2} + \frac{\left(e^{2y} - 1\right)^2\left(e^{2y} - 2y - 1\right)}{e^{6y} + e^{2y}(7 - 4y) - 2y + e^{4y}\left(-5 + 6y - 8y^2\right) - 3}, \tag{B.3a}$$

$$\frac{C_{\mathrm{p}l}}{Nk_{\mathrm{B}}} = \frac{3}{2} + \frac{\left(e^{2y} - 1\right)^2\left[1 + e^{2y}(2y - 1)\right]}{1 + e^{6y}(2y - 3) + e^{4y}(4y + 7) - e^{2y}\left(8y^2 + 6y + 5\right)}, \tag{B.3b}$$

$$\begin{aligned}\frac{\Delta C_{\mathrm{p}}}{Nk_{\mathrm{B}}} &\equiv \frac{C_{\mathrm{p}l} - C_{\mathrm{pg}}}{Nk_{\mathrm{B}}} \\ &= \left(e^{2y} - 1\right)^2\left[\frac{1 - e^{2y} + 2y}{e^{6y} + e^{2y}(7 - 4y) - 2y - 3 + e^{4y}\left(-5 + 6y - 8y^2\right)}\right. \\ &\quad \left. + \frac{1 + e^{2y}(-1 + 2y)}{1 + e^{6y}(-3 + 2y) + e^{4y}(7 + 4y) - e^{2y}\left(5 + 6y + 8y^2\right)}\right].\end{aligned} \tag{B.3c}$$

Appendix C

Formula for the heat capacity at constant volume in the coexistence region along the critical isochore

In terms of Lekner's parameter y, the expression for $C_V(y)/(Nk_B)$ for $\hat{V} = 1$ is obtained from equations (7.21) as

$$\frac{C_V(y)}{Nk_B} = \frac{C_V^{(1)}}{C_V^{(2)}},$$

$$C_V^{(1)} = \mathrm{csch}(y)\Big[2\big(-167 + 950y^2 - 1952y^4 + 768y^6\big)\cosh(y)$$
$$+ 4\big(158 - 567y^2 + 648y^4\big)\cosh(3y) - \big(400 + 20y^2 - 1248y^4\big)\cosh(5y)$$
$$+ \big(113 + 388y^2 + 64y^4\big)\cosh(7y) - 11\cosh(9y)$$
$$- \big(2840y - 2400y^3 + 3584y^5\big)\sinh(y) + \big(804y + 2480y^3 - 1920y^5\big)\sinh(3y)$$
$$+ \big(356y - 1520y^3 - 128y^5\big)\sinh(5y) - \big(206y + 320y^3\big)\sinh(7y)$$
$$+ 10y\sinh(9y)\Big],$$

$$C_V^{(2)} = 8[-2y + \sinh(2y)]\Big[42 + 12y^2 + 32y^4 - \big(61 + 48y^2\big)\cosh(2y)$$
$$+ \big(22 + 36y^2\big)\cosh(4y) - 3\cosh(6y) + \big(42 - 32y^3\big)\sinh(2y)$$
$$- \big(24 + 16y^3\big)\sinh(4y) + 2y\sinh(6y)\Big]. \qquad \text{(C.1)}$$

To calculate $C_V/(Nk_B)$ versus $\hat{\tau}$, the above quantity $C_V/(Nk_B)$ is calculated versus y and $\hat{\tau} \equiv \hat{\tau}_X$ in equation (7.3*b*) is also calculated versus y. Then one obtains $C_V/(Nk_B)$ versus $\hat{\tau}$ with y as an implicit parameter.

doi:10.1088/978-1-627-05532-1ch14

◎ 编辑手记

本书是一部英文原版的大学物理教材，中文书名或可译为《范德瓦尔斯流体热力学的进展》.

本书的作者是大卫·C.约翰斯顿(David C. Johnston)，目前任职于爱荷华州立大学物理学与天文学系和艾姆斯实验室.

大卫·C.约翰斯顿博士是爱荷华州埃姆斯市爱荷华州立大学物理学与天文学系的杰出教授.他分别在加利福尼亚大学圣塔芭芭拉分校和加利福尼亚大学圣地亚哥分校获得物理学学士学位和博士学位.在加入爱荷华州立大学之前，他在新泽西州安嫩代尔的埃克森研究与工程公司的研究实验室进行研究.他的研究领域是实验固体物理学，重点研究固体的电子、磁、热能和超导性质的测量和理论建模.他是美国物理学会的会员，也做过《物理评论快讯》(*Physical Review Letters*)杂志的部门副主编.

据本书作者自己在前言中所介绍的：

> 这本书是我在爱荷华州立大学教授的热力学和统计力学课程的基础上发展而来的.本书涵盖的主题之一是相变，包含的相变类型之一是流体的范德瓦尔斯平均场理论所描述的一阶液体到气体的转变，该理论由范德瓦尔斯(van der Waals)于1873年提出，并由麦克斯韦(Maxwell)于1875年扩充，是第一个根据粒子之间的相互作用预测相变的理论.但是，范德瓦尔斯状态方程和自由能导出的流体(气体或液体)的热力学性质这两方面目前尚未得到充分研究.

本书全面地介绍了范德瓦尔斯流体的热力学性质,其中包括对过去工作的回顾以及对我本人近期广泛研究的介绍.该书的主要目的是通过阐述范德瓦尔斯流体的许多有趣且多样化的热力学性质,以及描述它的各种具有独立参数的图表,以图形方式概述范德瓦尔斯流体的特性.这些图的数据是从本书导出的公式获得的,其中一些公式先前已出现在其他文献中,本书还包括许多不适用于图形显示的结果.

我希望这本书对讲师而言是有用的教学资源,对学生而言可以作为热力学和统计力学课程的文字补充,并且对于对范德瓦尔斯流体的热力学感兴趣的其他人也有一定的用处.

本书的内容还是相对比较浅显的,我们从目录中即可看出.

国外的大学教科书有两个极端,要么大而全,洋洋上千页,售价合人民币几千元,让学生不堪重负,要么简单明快直奔主题,不拖泥带水.有个段子:

> 据说,有几个美国宇航员发现了一个大问题,他们带的圆珠笔到了太空由于没有重力,不出水写不了字,因此无法记录.于是,美国人接下来花了好几年时间,研发了一款能够摆脱重力正常书写的圆珠笔.结果,他们惊讶地发现,那些俄罗斯宇航员只需要带着铅笔就可以了.
>
> 这个段子是查理·芒格(Cyharlie Thomas Munger)讲的,这位巴菲特的黄金搭档想通过这个段子说明,"千万要排除不需要的信息,别把事情搞复杂,记住你原来要做的事."

不像我们在评职或考核工作量时,规定教材要多少字,主编、副主编、编委挂一群人,还要分字数,所以中庸的居多.当然最后大多沦落为鸡肋.负能量打住!

刘培杰

2021.2.2

于哈工大

刘培杰物理工作室
已出版(即将出版)图书目录

序号	书　　名	出版时间	定　价
1	物理学中的几何方法	2017—06	88.00
2	量子力学原理.上	2016—01	38.00
3	时标动力学方程的指数型二分性与周期解	2016—04	48.00
4	重刚体绕不动点运动方程的积分法	2016—05	68.00
5	水轮机水力稳定性	2016—05	48.00
6	Lévy 噪音驱动的传染病模型的动力学行为	2016—05	48.00
7	铣加工动力学系统稳定性研究的数学方法	2016—11	28.00
8	粒子图像测速仪实用指南:第二版	2017—08	78.00
9	锥形波入射粗糙表面反散射问题理论与算法	2018—03	68.00
10	混沌动力学:分形、平铺、代换	2019—09	48.00
11	从开普勒到阿诺德——三体问题的历史	2014—05	298.00
12	数学物理大百科全书.第 1 卷	2016—01	418.00
13	数学物理大百科全书.第 2 卷	2016—01	408.00
14	数学物理大百科全书.第 3 卷	2016—01	396.00
15	数学物理大百科全书.第 4 卷	2016—01	408.00
16	数学物理大百科全书.第 5 卷	2016—01	368.00
17	量子机器学习中数据挖掘的量子计算方法	2016—01	98.00
18	量子物理的非常规方法	2016—01	118.00
19	运输过程的统一非局部理论:广义波尔兹曼物理动力学,第 2 版	2016—01	198.00
20	量子力学与经典力学之间的联系在原子、分子及电动力学系统建模中的应用	2016—01	58.00
21	动力系统与统计力学:英文	2018—09	118.00
22	表示论与动力系统:英文	2018—09	118.00
23	工程师与科学家微分方程用书:第 4 版	2019—07	58.00
24	工程师与科学家统计学:第 4 版	2019—06	58.00
25	通往天文学的途径:第 5 版	2019—05	58.00
26	量子世界中的蝴蝶:最迷人的量子分形故事	2020—06	118.00
27	走进量子力学	2020—06	118.00
28	计算物理学概论	2020—06	48.00
29	物质,空间和时间的理论:量子理论	2020—10	48.00
30	物质,空间和时间的理论:经典理论	2020—10	48.00
31	量子场理论:解释世界的神秘背景	2020—07	38.00
32	计算物理学概论	2020—06	48.00
33	行星状星云	2020—10	38.00

刘培杰物理工作室
已出版(即将出版)图书目录

序号	书　　名	出版时间	定　价
34	基本宇宙学:从亚里士多德的宇宙到大爆炸	2020－08	58.00
35	数学磁流体力学	2020－07	58.00
36	高考物理解题金典(第 2 版)	2019－05	68.00
37	高考物理压轴题全解	2017－04	48.00
38	高中物理经典问题 25 讲	2017－05	28.00
39	高中物理教学讲义	2018－01	48.00
40	1000 个国外中学物理好题	2012－04	48.00
41	数学解题中的物理方法	2011－06	28.00
42	力学在几何中的一些应用	2013－01	38.00
43	物理奥林匹克竞赛大题典——力学卷	2014－11	48.00
44	物理奥林匹克竞赛大题典——热学卷	2014－04	28.00
45	物理奥林匹克竞赛大题典——电磁学卷	2015－07	48.00
46	物理奥林匹克竞赛大题典——光学与近代物理卷	2014－06	28.00
47	电磁理论	2020－08	48.00
48	连续介质力学中的非线性问题	2020－09	78.00
49	力学若干基本问题的发展概论	2020－11	48.00
50	狭义相对论与广义相对论:时空与引力导论(英文)	2021－07	88.00
51	束流物理学和粒子加速器的实践介绍:第 2 版(英文)	2021－07	88.00
52	凝聚态物理中的拓扑和微分几何简介(英文)	2021－05	88.00
53	广义相对论:黑洞、引力波和宇宙学介绍(英文)	2021－06	68.00
54	现代分析电磁均质化(英文)	2021－06	68.00
55	为科学家提供的基本流体动力学(英文)	2021－06	88.00
56	视觉天文学:理解夜空的指南(英文)	2021－06	68.00
57	物理学中的计算方法(英文)	2021－06	68.00
58	单星的结构与演化:导论(英文)	2021－06	108.00
59	超越居里:1903 年至 1963 年物理界四位女性及其著名发现(英文)	2021－06	68.00
60	范德瓦尔斯流体热力学的进展(英文)	2021－06	68.00
61	先进的托卡马克稳定性理论(英文)	2021－06	88.00
62	经典场论导论:基本相互作用的过程(英文)	2021－07	88.00
63	光致电离量子动力学方法原理(英文)	2021－07	108.00
64	经典域论和应力:能量张量(英文)	2021－05	88.00

联系地址:哈尔滨市南岗区复华四道街 10 号　**哈尔滨工业大学出版社刘培杰物理工作室**
网　　址:http://lpj.hit.edu.cn/
邮　　编:150006
联系电话:0451－86281378　　13904613167
E-mail:lpj1378@163.com